The Executive Guide to Breakthrough Project Management

Capital & construction projects; on-time in less time; on-budget at lower cost; without compromise.

By Ian Heptinstall & Robert Bolton

Denehurst Publishing
Northwich, UK
2016

www.BreakthroughProjectManagement.com

First Edition 2016

Published by Denehurst Publishing,

An imprint of Denehurst Consulting Limited
Northwich, CW8 2XH
United Kingdom

ISBN: 978-0-9954876-0-4 (paperback)
ISBN: 978-0-9954876-1-1 (e-book)

E: contact@BreakthroughProjectManagement.com

W: www.BreakthroughProjectManagement.com

Contents

Preface

In this book, we recommend that changes are made to how capital and construction projects are planned and managed.

This is not change for change's sake.

Whether you are a project client or a member of the project supply chain, the techniques that you will learn about in this book, will deliver significant and sustainable improvements in the profitability of your business.

The methods underpinning the changes we suggest are well-established, having been developed in the 1990's. Whilst they have proven themselves, they have not yet become "mainstream" in the project management profession. Not because they don't work, but because changing established habits and practices takes time, and meets with resistance.

Most projects use the same well-established methods for contracting, planning and managing, without realising that there are better alternatives.

Having used these methods ourselves, we

were puzzled at why they are not more widely used on capex and construction projects. We have seen the significant improvements they can bring. We are not talking about small changes, we are talking about projects delivered in over 30% less time, and at least 15% lower cost.

As part of our research for this book, we talked with project managers who knew about these methods, about why their application is so limited. There was a common theme in what they told us:

- "They won't let us",
- "They insist we do it this way",
- "They give the project manager autonomy to manage how they want to",
- "They have just spent a fortune on a new IT system/training/ qualifications/consultants so they want to stick with that for now".

The finger is pointed at 'them', the leaders of the organisations who commission projects, the major contractors, and even investors and the providers of capital funding.

We don't believe that these senior executives are deliberately holding their organisations back, nor do we believe that project clients want to spend more money on a project, or to have it complete much later, with a lower quality outcome.

So we wrote this book. We want to address the problem that "I didn't know".

The need to do this is becoming more and more important. Growing populations need more infrastructure, but at the same time there is pressure on budgets, and a need to get "more from less". As the business world becomes more and more volatile, and dramatic changes seem to happen overnight, the days of long-term investment paybacks seem to be numbered – investors are demanding shorter investment cycles, and lower levels of investment.

Although the construction industry has struggled for over 40 years to improve productivity, we believe that those who successfully adopt the methods we describe in this book, will notice an immediate and substantial improvement in project productivity.

There is nothing in this book that is beyond the capabilities of most organisations ... if they are willing to change their current practices and habits. The techniques the you will learn can be used across a wide range of project types and sizes, whether they are worth less than £1 million or over £1 billion.

However, changing ingrained habits is not an easy thing to do, and without the active support of Senior Executives, change is unlikely to stick.

Senior-level leadership is key to making the changes that we recommend.

And what better time to do this that now? The world is crying out for innovators and leaders to build on the lessons learned by previous generations, and make their mark on an industry that has struggled to find a repeatable way to improve its performance.

We hope you find the book interesting, stimulating, and most importantly, that it encourages you to "give it a go".

Ian Heptinstall   Robert Bolton

contact@BreakthroughProjectManagement.com

Authors' Note:

Throughout this book we will often use the term 'contractor' to mean the external organisations that have the greatest impact on a project's success. It is a broader definition that the more common usage.

This includes architects, designers & design consultants, main contractors, specialist sub-contractors, manufacturers of significant equipment, large wholesalers or distributors, and maybe even a project management consultant.

Introduction

In 1965 the American high jumper Dick Fosbury introduced the world to a new way of jumping. Although it sounded crazy – to jump head first, with you back to the bar - he had science on his side.

Even when the success of the technique became plain to see, many athletes stuck with what they knew best, and some of the more skilled jumpers continued to win competitions... for a short while.

But there was no turning back, and since 1977 every world record holder has used the "Fosbury Flop", as this technique became known.

We have written this book, to introduce the capex project industry to the "Fosbury Flop of project management".

We believe that in the future, we will see that those who deliver the fastest and lowest cost projects are using the methods that you will learn in this book.

Breakthrough Project Management has

been written for Capital & Construction projects, those that involve major investment in assets and infrastructure, such as offices, hospitals, roads, rail systems, factories, and production plant.

These type of projects go by many names, including construction, infrastructure, mega-projects, capital projects, CapEx (capital expenditure), engineering, and EPC (Engineer-Procure-Construct).

For simplicity we will refer to them as capex projects.

The distinguishing feature of this kind of project is that the majority of the work is done by contracted suppliers and contractors, rather than employees of the client organisation. Many clients even outsource the overall project management responsibility, often without any understanding of how projects should and could be managed, and without realising the significant impact of the decisions that are being made on their behalf.

Outsourcing the management of a capex project is not itself a problem. However, the way in which most work on projects is outsourced, in our view, is a major problem, one that lies at the heart of many of the issues that beset capex projects.

The most common types of contract used to engage the key members of the project team, reward the project's suppliers, contractors and consultants for looking after their own interests, rather than focusing on making the project more successful for the client. This introduces a significant conflict in most project teams – should I do what will help the overall project to succeed, or should I do what will help my employer make more profit? When projects do not go smoothly, you can't do both, it is an either/or choice. It is "me" versus "we", and this dilemma can destroy the kind of high-performing teamwork that is necessary for all but the most trivial of projects.

For decades, report after report has highlighted the low levels of achievement of not only capex projects, but projects of all kinds. Far too many fail to achieve their objectives, finish late, or spend more than was budgeted.

In this Executive Guide you will learn an approach to getting away from the poor project performance levels that are so prevalent today.

We are not going to just suggest that you work harder, do more training, or use better people. No, we want you to "do it differently". We believe that the root cause of most of today's project problems is **how** capex projects are managed and procured, rather than **who** manages them. The issue is the system, not the people.

Organisations will not escape from poor project performance until they change how projects and their procurement are managed. And in this book you will learn how combining two innovative and proven techniques offers a reliable and repeatable way to bring about this change.

These innovations offer a robust and sustainable way for capex projects to reduce risk, reduce cost and reduce the duration of projects.

No longer is it a case of; *"Time, cost, or scope. Choose any two"*. Now you can have all three!

But not without challenging your understanding of what represents best practice in project and contract management.

This Executive Guide is a brief introduction to the ideas of Breakthrough Project Management, offering a route to significant improvements in business:

- Increased return on investment for project owners and investors
- Improved profitability for contractors, despite lower project costs
- Reduced risk, with improved visibility of emerging issues and risks – so that they can be tackled when they are small and can be easily rectified
- More projects for the same investment and with the same resources – or the same projects delivered using less resource
- A higher quality outcome
- Fewer disputes and faster final-account settlement
- Reduced levels of stress across project team members, and the ability to attract the best employees and suppliers
- Contractors can still win lowest-price bids, and deliver a great result at higher profitability

Working in the way we propose is not complicated, but it is different from the methods used to manage the majority of capex projects and project portfolios.

Proven to deliver step change project performance improvement in many different industries, the two fundamental pillars of Breakthrough PM are:

1. Plan and manage your project using **CCPM (Critical Chain Project Management)**.

   Chapter 2 explains how CCPM works, and gives examples of organisations who have used it to deliver projects on-time in much less time

2. Select and contract with your most important project contractors/suppliers, using a **Project Alliance** to align the interests of all team members.

   Chapter 3 discusses collaborative contracting approaches, and how they deliver better performance at lower cost – a Project Alliance is a form of collaborative contract.

Breakthrough PM requires both these elements. One without the other can be made to work, but will be challenging, and require extensive management time. It may deliver some improvement in projects' performance, but nowhere near the consistent step change in performance that is possible by using both pillars together.

Implementing Breakthrough PM does not change every aspect of how you manage projects.

Whilst it involves a significant change in scheduling, progress management, and procurement, it is completely compatible, and in many case enhances, a wide range of other approaches to improving project performance. Chapter 4 lists several value-enhancing techniques used on capex projects, and discusses their compatibility, or incompatibility, with Breakthrough PM.

The final chapter – chapter 5 – outlines the key factors in successfully implementing Breakthrough PM in a range of different types of organisation.

We hope that you find this short book thought-provoking, and look forward to hearing from you as these concepts get put into practice.

You can find out more, discuss your thoughts with us and with other industry professionals, as well as engage in discussing the ideas we outline in this book at our website.

We look forward to seeing you there.

www.BreakthroughProjectManagement.com

1 The Need for Change

"We cannot solve our problems with the same thinking we used when we created them."

"Insanity: doing the same thing over and over again and expecting different results."

Albert Einstein

The Current System is Broken

The current methods used to manage projects are not good enough. There is too much evidence that shows that even when projects rigorously apply the accepted methods of project management, they can still be late and still cost more than budgeted.

By 'accepted methods' we mean (i) the use of critical path or sequenced tasks, to plan; (ii) managing progress by focusing on tasks and milestones, and driving them to complete on their planned completion date, maybe using techniques such as earned value management (EVM)[1]; and (iii) contracting and subcontracting using fixed-price contracts wherever possible.

In practice, good project performance usually comes from a good project manager instinctively knowing what to do. Or luck.

A quick Internet search for 'major project failures' will provide sufficient supporting evidence for our assertion. A study of over 350 Oil and Gas mega-projects published in 2014 said that 64% of them overspent and 73% took longer than planned (Ernst & Young, 2014). Data from 2012, covering all capex project types showed remarkably similar results; 63% of projects are over budget, and 75% are late (AT Kearney, 2012). A recent McKinsey review of mega-projects claims that a staggering 98% of projects suffer cost overruns of at least 30%, and 77% are at least 40% late. (McKinsey, 2015).

This is despite the use of well-established software tools, structured and well documented methodologies, and more project managers with formal qualifications in project management. It seems that none of these factors is a reliable predictor of great project performance.

If the current project management methods were the best way to manage projects, then there wouldn't be so many problems with major projects, and we could reasonably expect to see a gradual improvement over the years as

---

[1] If you use EVM, you might be interested in one of the papers on the book website, comparing EVM and CCPM.

'best practice' becomes established and more and more project managers become qualified.

But we don't see continuous improvement; in fact, it seems to be just the opposite.

Project management author and consultant Lawrence Leach, in his recently updated book (Leach, 2014) reports that this situation has not changed much over the past 20-30 years – most projects fail to achieve their original time, cost & scope objectives.

Paul Teicholz at Stanford University, and Matt Stevens from the University of Melbourne, have highlighted, in separate studies, a steady decline in construction productivity in the USA since 1964, whilst productivity in all other sectors have significantly increased (Lean Construction Institute, 2014). We do not believe that other countries are much different.

Massive Improvement is Possible

The two pillars of Breakthrough PM have been proven over-and-over again to deliver sustainable improvement in projects.

Critical Chain Project Management (CCPM)

CCPM was developed in the mid-1990's, and first came to prominence through the book *Critical Chain* (Goldratt, 1997). Since then, organisations large and small have publically shared their results, confirming that it works.

When correctly implemented in organisations that understand the basics of project management, it delivers:

- Shorter durations – a CCPM project is usually at least 30% shorter
- Lower use of resources – Organisations can deliver 50-100% more projects, without increasing resources
- Lower costs – through using less time-related resources. If half of your project's cost are time-related (for example, people and hired equipment), then the time reduction alone will reduce cost by 15% on a typical CCPM project
- More reliable completion dates – typically well over 90% on-time completion (and with a shorter overall duration)
- Lower levels of overwork and stress across the team

Collaborative Contracting

Industry has known for several decades that the adversarial nature of the contracts used on capex projects has a negative impact on performance. Collaborative/ Relational Contracting has been proposed as a much more effective approach. Well known support for the idea includes the UK's Latham & Egan Reports produced in the 1990's.

Also in the early 1990's the US Construction Industries Institute investigated the benefits coming from what was known as 'project partnering'. Their research into almost 300 projects (CII, 1996), reported that collaborative project teams typically delivered projects:

- 20% faster

- 10% cheaper
- With fewer injuries and safety incidents
- With improved client satisfaction
- With higher supply chain profitability

Whilst there are a number of ways in which the members of a project team can be selected and contracted collaboratively, Breakthrough PM recommends the approach of using a *Project Alliance*, a well-established form of collaborative contracting.

Breakthrough Project Management

Breakthrough PM combines these two approaches. We believe that this gives a more robust and repeatable model than operating either approach by itself on a capex project.

A collaborative team by itself does not automatically deliver improvement. You don't just put people together, remove commercial barriers to collaboration, and hope they magically deliver a better project. It's possible, but it isn't guaranteed.

Adding CCPM to a collaborative project team provides a mechanism to deliver improvement, and it embeds collaborative behaviours into everyone's day-to-day work.

CCPM requires a collaborative project team in order to be successful. On capex projects, most of the project team is not employed by the project client or owner, there is a commercial contract (or chain of contracts) between client and team member. The common contracting method of using fixed-price contracting with the project suppliers/contractors, who are often selected based on lowest bid price, makes CCPM difficult, if not impossible, to implement.

Even if the majority of your current projects do achieve their targets, if you are using conventional project management methods, then your project durations are probably longer than they need to be, and this means that they will be costing you more than they need to.

A Competitive Market Doesn't Guarantee Excellence

Market competition does not automatically lead to suppliers seeking out and operating according to so-called 'best practice'. To survive in a competitive market, you only need to be as good as your significant competitors. This often means being good enough to win enough business to keep you going.

Most of us can hold our own in a competitive 100-metre race amongst our friends. If we all eat and drink a little too much, and don't do enough exercise, there would still be a competitive race, and there would still be some variation in who wins. But that doesn't make us the best we could be, and it leaves our winning times a long way away from Usain Bolt's current world record of 9.58 seconds. Even if I (Ian) wanted to win every race, I only need to be a few percent better than the others in the race. I don't need to be any better than that.

But if Robert hired an Olympic-level coach, followed the right diet, trained rigorously, what do you think would happen? He could probably win at will, without breaking sweat, and if I wanted to win again, I couldn't do so overnight (if at all).

Bidding and using market competition works in exactly the same way. Suppliers only need to be as good as their competition. And since almost no one is using the methods in this book in the way we recommend, then there are no market forces pushing for change. Everyone does it the same way, using the same pool of people, and similar methods. Bidders are happy to squeeze their margins a few percent when they need some more work. They don't need to do more than that.

Until that is, one company decides to be different, and by the time their competitors notice, they could be a long way ahead.

The Risks of Inaction

Are you willing to take the same risk that high jumpers, back in the late 1960's and early 1970's, took when they stuck with the tried-and-tested straddle jump, despite Dick Fosbery's demonstration of a 'better way'?

Companies large and small are using the principles behind Breakthrough PM today, in industries ranging from software to new product development and major maintenance.

But in the capex industry, their use is rare. We discuss our views on why this might be in later chapters, but it is not because the methods don't work.

The relatively low take up today offers a once-in-a-lifetime opportunity for early adopters to develop what INSEAD professors Kim and Mauborgne call a "Blue Ocean Strategy" – ie a decisive competitive edge with few competitors[2].

There are project staff in most major contractors who are aware of the ideas, but most clients insist on following convention, so to the contractor, the ideas are little more than an interesting theory.

Clients are not demanding their suppliers change and improve the way that they manage projects. They either have their own views on how projects should be managed based on current orthodoxy, or they assume market competition automatically drives innovation.

Even when clients leave their supply base to manage projects in their preferred way, most contractors seem to be happy to work in the same way they always have, and in the way most of their competitors do.

As we write this book, the Project Management Institute (PMI) has just published their 2016 "Pulse of the Profession" survey. Over a third of respondents said that they used CCPM 'always' or 'often', and it seems that outside of construction CCPM is moving from a niche idea, to a mainstream method for managing projects[3].

What if one of your significant competitors tries the ideas in this book, and what if they succeed in making them work? What if you are a contractor and an important

---

[2] *"Blue Ocean Strategy"* by W. Chan Kim and Renée Mauborgne, 2005/2015 Harvard Business Review Press.
[3] CCPM had the same level of 'always' and 'often' adoption to Agile, and higher levels that Lean, Scrum & Six-Sigma PM techniques. See p21 in PMI's Pulse of the Profession® 2016,

client starts to ask for proposals following these principles? Do you really want to have to learn at the same time as putting a tender together?

It will only take one competitor to use these principles to have a significant impact on your business. It could even be a company that you consider to be too small to be a competitive threat to you.

By the time you notice that they are doing something different, you could be years behind. We don't think there is enough profitability in the industry for you to keep operating as you do today, to squeeze margins in order to win new work against the new competition, and also to catch up. Unless you have very deep pockets...

What's in it for me – as a Client?

- Better ROI from your projects
- Shorter projects – positive income and breakeven both happen sooner
- Lower project costs
- Cash freed up to do more projects
- Improved quality of the final project product
- Less risk that contractors have to cut corners to make money
- Lower levels of project risk

- As a more attractive customer, you ensure the best contractors are interested in your bid, reducing the number of no-bids, or over-priced bids[4].

What's in it for me – as a Contractor?

- Improved business profitability
- Sales increase, without increasing cost or overhead, and making better use of key resources
- Reduced cashflow risk
- Higher reputation
- Lower financial exposure from problem projects
- Staff retention

The Focus of this Book

Breakthrough PM addresses the parts of project management convention that we believe are holding projects back, and need to change. In particular, we address scheduling, project control and execution management, the procurement of the major project contractors and suppliers, and programme/portfolio management.

We are not claiming that it is an all-encompassing body of knowledge for project management. You still need to define the right project, and to integrate good practices in risk management, continuous improvement, quality control, and lean principles. You still need to use the right

---

[4] Rather than submit a no-bid, contractors will often bid a high price, not expecting to win. This is because they worry they might be excluded from future bidder lists if they don't submit a bid.

tools to design and communicate across the team. You still need good project leaders.

Breakthrough PM provides a great foundation for a project team that is better placed to use and exploit tried and trusted good practices such as risk and value management, scope definition assessment (for example using tools such as PDRI[5]). The use of collaborative design and knowledge management tools (such as BIM[6] and PDMS[7]) are fully compatible with Breakthrough PM.

In writing this book, our focus has been just on project execution and delivery, and unlocking the significant improvement potential. The main reasons for this are the lack of published material on the ideas, and the size and speed of the improvement that adopters can realise.

Once mastered, the underlying philosophies behind Breakthrough PM can also help in project selection and strategy development, as well as helping to ensure the right projects are chosen and specified, ahead of execution.

If you would like to learn more about the applications that extend beyond the scope of this short guide, we'd be delighted to hear from you.

---

5 *Project Definition Rating Index (PDRI), a tool to measure the quality of project definition, developed by the Construction Industries Institute (www.construction-institute.org)*
6 *Building Information Management. 3D design tool with associated detailed database, used in the buildings construction industry.*
7 *Plant Design Management System. Similar to BIM - used more in the process industries.*

The Executive Guide to
Breakthrough Project Management

2 | Manage Projects using CCPM

"If you agree with every step of the argument, but the conclusion leaves you angry or uncomfortable, it might be time to reconsider your world view, not reject the argument."

Seth Godin

CCPM in practice:

• Emesa supplied the station structure for the new TVG station in Liege Belgium – a €50M, 3-year contract. With only 6 months to go, they were 5 months behind schedule!   They implemented CCPM, which helped them deliver on time and avoid €5 million in late penalties. *In effect they delivered 11 months' traditional work, in 6.*

• When Boeing introduced CCPM into its airframe design process, it completed the design in *30% less time than the previous best. They recovered a 2-month delay in starting, and had 50% fewer errors.*

• Japanese construction company

Daiwa House used CCPM to turn around a failing ERP system implementation. It was 4 months behind schedule after just 13 months. 12 months later the project was completed on time, and used **27% *less* *implementation resources*** than planned, *saving over $10M* in external consultant fees.

- A joint-venture led by Primex was one of three contractors building 100km of road in Mexico. Within a few months unexpected ground conditions meant they were 45 days behind. Using CCPM, they *completed on time, whilst the other two contractors (larger, more experienced, and not using CCPM) completed their sections 40% late*

The difference that project planning and execution management can make is of critical strategic importance.

What would be the impact on your business if you delivered the kind of improvements the above companies achieved? These results are typical for CCPM implementations, with many reducing project durations by over 50% compared to traditional methods.

The premise of Breakthrough PM is that the conventional ways in which projects today are planned and executed is the root cause of most delivery problems. It is why so many projects have poor performance, despite having industry veterans, qualified staff, and expensive software. Changing this is the key to delivering a project faster, at lower cost, and with much greater predictability.

Breakthrough PM uses Critical Chain Project Management (CCPM) to schedule the project, and to manage its execution.

> Amdocs, a $4B turnover B2B software company, put large numbers of their project managers through PMI qualification programmes because they thought that their project productivity was a function of the skill levels of their team.
>
> This made no difference to the productivity or quality in design, nor to project delivery.
>
> Within months of starting to use CCPM they noticed improvements.
>
> In less than one year, all their projects used CCPM, and Amdocs were delivering 14% more projects with the same resources, and completing projects in 20% less time. Crises requiring main board intervention now rarely happen, costs are down, cost overruns are rare, and they can more easily cope with late client-driven changes.
>
> *Source: Discussion with Yoav Ziv, VP of Amdocs, February 2015*

Developed in the 1990's, CCPM has been used in thousands of projects across the world. However, it is still in its infancy, and awareness of CCPM varies widely across the project world. There is also a lot of misunderstanding of CCPM amongst many project management experts, with one of the world's leading CCPM practitioners saying that misunderstanding of the CCPM methodology was 'prevalent' amongst project managers who had heard about it, but hadn't used it![8]

Having been involved in, and studied, hundreds of projects that have used CCPM; we know of no other

---

[8] Correspondence with a manager responsible for implementing CCPM at a global airline, early 2016.

structured method for managing projects that consistently delivers significant and rapid improvement, across a wide range of project environments.

CCPM incorporates many of the practices that the best project managers instinctively use, but unlike 'experience', it is teachable, and can be embedded into systems and simple processes. Results can be achieved quickly, and by junior staff just as well as those with more experience.

Some companies who use CCPM keep it quiet, because they see it as a competitive advantage. Some of the major organisations that have implemented CCPM, and publically shared the dramatic results include:

- Boeing (designing & building airplanes)

- Amdocs (software). See sidebar

- Seagate (digital memory)

- Lufthansa (aircraft maintenance support

- Harris Semiconductor (new factory)

- Siemens (power stations)

- Tata (steel manufacturing shutdowns)

- Balfour Beatty (construction – in a 1990's pilot)

- Delta Airlines (jet engine overhaul, aircraft maintenance)

- French & US military (repair & overhaul of equipment)

- Japan's Ministry of Land Infrastructure & Tourism (public infrastructure)

- Metodo Engenharia (Brazilian contractor)

- Mazda (new technology development)

- Unilever (new factory build)

- NASA

CCPM is not just an alternative way to plan and manage a project. It delivers improved results. Most of these companies had extensive experience in managing projects before they tried CCPM, and they had well trained staff supported by sophisticated project management software. They still found that CCPM increased their performance enough for them to go public with the results. A quick look at the websites of the main CCPM software companies produces a list of hundreds of companies using CCPM, in all imaginable project environments, and pages of detailed case studies, demonstrating step-change improvements similar to those we outline here.[9]

What makes CCPM different?

One of the core reasons that current project management methods fail to deliver the results that businesses need, is in the way they address uncertainty.

Projects are inherently uncertain, and there are many unknowns in putting a schedule and budget together. If you look at the research into why projects fail, you will see that many of the reasons given cite uncertainty. To quote just one report "...most clients would say they were unforeseen failures with few or no warning signs to help avoid disastrous results." (KPMG, 2013).

We don't fully agree with the implication in this quotation, that the impact on projects of this uncertainty is inevitable.

---

[9] In October 2015, we found over 250 companies listed, and over 70 case studies on just 4 websites: www.exepron.com, www.realization.com, www.prochain.com, www.beingmanagement.com.

We all know something will happen on projects to make some tasks more difficult than expected, whilst others will go more smoothly. The problem is that we don't know ahead of time which tasks will go smoothly, and which will have problems.

The traditional approach to project management expects each task to manage its own uncertainty, and to make sure it is delivered on time. The idea is that if each task can finish on time, then the whole project will finish on time. When major elements of the project are sub-contracted, this philosophy is built into the contracts, and reinforced with penalties or liquidated damages if tasks are late.

CCPM takes a fundamentally different approach to managing the inherent uncertainty that exists in projects. It is far too important to leave it to the individual task owners.

Uncertainty is part of life, and something that we all understand intuitively. The problem is we rarely apply our real-world common sense when we manage projects.

Imagine you live in a busy city, 10 miles from your normal place of work. If you were asked how long it took you to travel home in the evening, you might give the average time – say 30 minutes. If you were asked what is the shortest time it has taken you to make the journey, you might say 18 minutes, and for the longest, 90 minutes.

We instinctively know that there are all kinds of things that can happen that are outside of our control, and which mean the actual time you need on any one particular day is unpredictable. In fact, the probability of it taking the exact average time (30 minutes precisely) is quite low.

But when it comes to projects, we don't like task managers to be honest, and saying "It will take somewhere between 18 and 90 minutes". It can sometimes seem that senior managers live in a much more predictable world than the rest of us (or to give them the benefit of the doubt, perhaps they genuinely believe that the best thing to do is to insist upon using a 'highly certain' number). Since being late on a task is seen as a bad thing, most of us, when forced to give a task duration, will give a number much larger than the average task duration.

Using our journey home as a simple example, if we had an important evening engagement, most of us would plan for much longer than the 30 minutes average; more like 60 minutes, perhaps 80 or even 90. We probably wouldn't plan to take 30 minutes if we wanted to be certain that we would not be late. We naturally add safety to the average task time in order to provide a reliable commitment.

This exact same phenomenon happens on projects. We ask task estimators to provide a reliable commitment, and then manage projects by expecting these commitments to be achieved. We make contracts with fixed-prices and invoke damages for late task completion, expecting each supplier or contractor to manage their own risk. This means that almost all the tasks are planned conservatively, even if the task owner is put under severe pressure to reduce their estimate. No one gives the planner an average time or cost, because by definition 'average' means that half of the time you will take longer and cost more, and to do so would be commercial suicide.

The safety built into these reliable commitments is not padding. It is often needed. It is necessary safety to handle the inherent uncertainty and variability in work.

In traditionally-managed projects this safety is included in task times. Often the estimator does not even realise that it is there – no one talks about 'average time plus safety'. Estimating norms are based on real experience, and thus include safety, because this cushion is often needed.

Although project managers do talk about 'contingency', this is typically a monetary value, to cover specific risk events, rather than natural variation around the mean. It is rare that contingency time is included in schedules. Contingency is also something that gets stripped from estimates to achieve budget targets, forcing wise estimators and planners to hide it.

There are some techniques, such as full PERT and Monte-Carlo simulation that try to account for the task time variability, but their use in real-world projects is rare. There are also some weaknesses in how they are implemented that convince us that the CCPM approach is both simpler and more reliable.

Building some safety into the duration of a single task is not significant by itself. However, across a project consisting of hundreds or thousands of such tasks it is an enormous issue – it means that there is far too much time included in the project plan.

As well as too much time (and therefore cost), trying to manage uncertainty with task-level safety also reduces the chance that the overall project benefits from an improvement in the duration of a single task. In general, early task finishes have little positive benefit on traditionally-managed projects, unlike in CCPM, which is designed to exploit early task completion.

In 2003 a small construction contractor in Japan piloted the use of CCPM on a flood-defence project.

Despite the project starting late, being subject to change, and having an inexperienced project leader, it was delivered early and under budget. The owner of Sunagogumi, the contractor, was convinced, and applied CCPM to all his contracts.

The success of project using CCPM came to the attention of the federal ministry, MLIT, who immediately saw its potential. Soon after their "Win:Win:Win Public Works" initiative was born, and CCPM became a cornerstone of Japan's national Construction Strategy.

Since then, thousands of ministry and contractor staff have learned and used CCPM to deliver infrastructure projects in Japan faster, better and cheaper.

Sources: Kishira (2009), & a meeting Ian had at MLIT Tokyo in Nov 2011, with Yuji Kishira and officials from MLIT.

How can CCPM schedule shorten projects, and still deliver on-time?

A typical CCPM project will remove at least 25% of the time included in a traditionally-planned schedule. And because many costs are proportional to the amount of time taken (for example the cost of people and hired equipment), then costs also come down. In addition,

CCPM-managed projects consistently finish on, or ahead of, schedule, compared with two-thirds of projects being late, reported by Ernst & Young (2014), and AT Kearney (2012).

But how can it do this? Experience tells us traditional projects are more often late than early, and this would imply there is not enough time in the schedule, yet CCPM is able to save both time and money.

It is all down to how the safety is used. CCPM is built on the premise that behaviours within a traditionally-managed project waste the built-in safety. CCPM includes the tools to ensure the safety time is very visible, and is managed carefully across the whole project. The key is in execution management, not just in the planning. CCPM doesn't just squeeze durations, if focuses on reducing wasted time.

With each task including an allowance of safety time and money[10] to cover the 50% chance that it will take longer than the average time, tasks will have more time allocated in the schedule than they will need in most circumstances. A typical project task estimate contains sufficient time to complete in about 90% of situations. In other words, in 89% of cases it will need less time than allowed.

If our experience tells us that many tasks finish late, and this in turn makes projects late, how can it be that if you reduce the estimated time for each task, this results in a better chance of finishing the overall project on time?

---

[10] For simplicity we will only use time safety, though the same principles apply to cost.

This is because many tasks will not need the safety that they have built in. Think about the previous example of your commute home. Whilst you may allow 60 minutes for the few occasions when arriving by a given time is essential (30 minutes average, plus 30 minutes safety), most of the time you will not need 60 minutes, and so you will arrive home early, or instead you might choose to take longer than necessary by stopping at a café, or pulling over and making some calls.

While this may not matter for your occasional drive home, consider the effect when this idea is multiplied across a project.

For example, we have a project that consists of 8 tasks, each of 4 weeks. Each task is done by a different supplier or contractor, and you have insisted on highly certain (fixed-price) bids. So they all include safety.

Task-level Safety	
Task Duration	4 weeks
Tasks	8
Project Duration	8 x 4 = 32 weeks

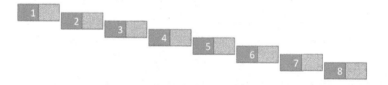

Let's say that the <u>average</u> task time is two weeks, with 2 weeks' safety – about 50% of the 'highly certain' estimate[11].

---

[11] This is a CCPM rule-of-thumb, and has been validated across thousands of projects and hundreds of thousands of tasks. Whilst there are exceptions it is a good-enough estimate.

If you were to take all of this safety out of the individual tasks, and put it into a pool to be used by those tasks that really need it, then this safety pool would need much less time than the total time you removed from each task.

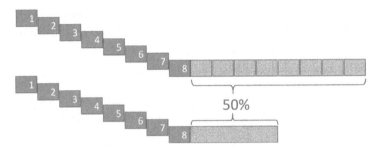

50%

In our 8 task example, a shared safety pool just 8 weeks, compared to the total of 16 weeks included in each task.

Pooled Safety	
Task Duration	2 weeks
Tasks	8
Removed safety	16 weeks (8 x 2 weeks)
Project safety pool	8 weeks (50% of the removed safety)
Project Duration	(8 x 2) + 8 = 24 weeks

The idea of risk (uncertainty) pooling is not new, and the fact that you need less safety in total if it is aggregated/pooled, can be proven mathematically if you are so inclined. Luckily this is an Executive Guide, so we will give this a miss, but the important point is that this is not magic, witchcraft, or just playing with numbers. This is a well proven principle, but a principle that is ignored in the way most projects are managed today.

Risk pooling has been used by the insurance industry for centuries. We each pay an amount to protect against a risk, into a centrally-managed pool. The amount we pay is much smaller than we would need to put aside if we had to self-insure. This is exactly what CCPM does on projects.

The first principle of CCPM is that uncertainty is managed, and provided for at the project-level, not the task-level. The time and cost allowance to cover against uncertainty and variability is aggregated. Tasks are sequenced, and their nominal duration is based on the average task time.

This flies in the face of the traditional practice. Traditional practice demands firm task completion dates, or milestones. It assumes that if we manage all the interim milestones closely, then the resulting project will be on time. It wants to be able to blame and penalise a particular contractor if they are late.

CCPM takes a different, and counter-intuitive approach. By ignoring interim milestones, and focusing intensely on the project completion, it means we monitor the consumption of the shared safety, and compare this against the overall project progress. Which tasks need to dip into the safety pool is not important.

Whilst risk pooling is central to CCPM's success, it is not sufficient just to pool the safety and then cut it. We also need to manage the project differently. Traditional project management techniques are designed to try and manage risk at the task level. Now we no longer have task-level safety, we need new management processes that are designed to manage pooled-safety. Without changing these practices, we will not get any improvement in performance.

In CCPM, the pooled safety is called a buffer. A project schedule will have a single project buffer, and a small number of additional strategically-placed buffers, that contain the safety. These are usually shown in a project bar chart as a 'task', as shown below.

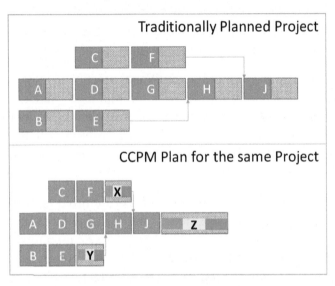

This is purely a schematic representation – the buffer is not a task in the traditional sense, and it is not used only at the end. The buffers are there to calculate the total project duration, and the necessary start dates for task chains in order to achieve the project end date.

In the illustration above, you will see that the traditional task durations have been halved, and a new final task (Z) called the 'project buffer' has been inserted. There are two other buffers, X and Y, added to ensure that the Critical

Chain (A-D-G-H-J) is protected from any overrun in the feeding chains C-F and B-E.[12]

Monitoring the consumption of the buffers is the main tool used for project execution management. Buffer consumption provides a simple and clear way to indicate priorities to task and resource managers, and at the same time give project managers and senior executives a simple snapshot of overall progress.

If we are 50% of the way through a project, and have used 30% of the project buffer, things are progressing well. If we have used 50% of the buffer, things are still OK, but we might want to start planning to recover some of it. If we have used 75% of the buffer, then we should already be implementing recovery plans.

The Fever Chart

Most projects use a simple graphical tool, called a fever chart, to monitor the project buffer. A simple fever chart is shown below.

---

[12] In this simple example the "Critical Chain" is the same as the "Critical Path". The main difference between a critical chain and a critical path is that the critical chain takes account of resource dependency and availability in the schedule, whereas the critical path is based on the logical sequence of the tasks. In order to keep this guide simple, we are not going into this difference, but on most real projects, the critical chain is not the same as the critical path.

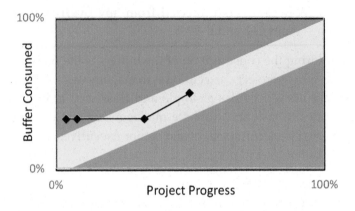

The points on the line are the progress reports. In reality, most projects would have many more points than shown here.

The chart shows that initial progress was slow, with about 33% of the buffer used before 10% of the project was completed, however by the third report, things were in balance – no more of the buffer had been used, and 33% of the project was complete. This progress continued up to the latest report, where we are now half complete, and have used 50% of the buffer. The colours give an immediate, and intuitive, indication of progress.

Green	**OK** - keep going
Yellow	**Watch out** - prepare recovery plans
Red	**Act** - implement recovery actions.

Using buffer management and a fever chart gives a better indication of progress than most established alternative techniques, and is particularly good at giving an early warning sign of potential problems. It overcomes several of the inherent problems in established project control methods such as Earned Value Management.

As well as being a tool used to manage an individual project, the fever chart also provides an excellent executive project portfolio dashboard, an example of which is shown below. In this graph the latest status for each project is represented by a single point.

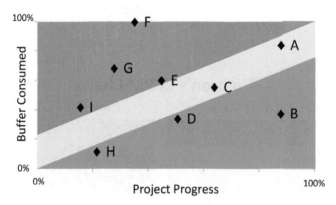

A reader can immediately see that projects F and G look to be the priorities for support or intervention, whilst B needs almost no attention.

An Alternative to the Fever Chart

Most CCPM software uses the diagonal style of fever chart, as in our images above. So if 30% of the buffer is used with only 10% progress, then buffer colour is red, indicating a problem.

One major CCPM software company – Exepron – uses a different approach, replacing the fever chart with two charts; the *Project Status* and the *Early Warning*.

The Project Status report simply shows how much of the buffer is used – green meaning < 33%, yellow 34-67%, and red meaning over 67% of the buffer has been used.

By itself this change would diminish one of CCPM's strong points, the ability to give a reliable early warning of emerging problems and delays, and to focus management's attention. This is why the developers of Exepron added the Early Warning chart, which considers a range of factors that can influence the risk to a project's on-time completion, not only buffer consumption.

This provides the management information in a different way, as shown below. So using our simple example, if 30% of the buffer was used with only 10% progress, whilst the buffer would be green, it is highly likely that the Early Warning chart would be showing black or red status (indicating a problem).

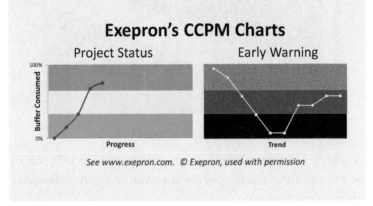

Exepron's CCPM Charts

See www.exepron.com. © Exepron, used with permission

CCPM's Key Principles

CCPM involves much more than we have included in this short book. The following section summarises the three core principles of CCPM, and outlines what they entail. We have included them to give a flavour of what CCPM entails, and to show that it involves more than just aggregating safety time.

There are many excellent text books on CCPM if you are interested in learning more about the method, and we have listed several in the Bibliography.

CCPM Principle 1: Buffers

The issues addressed:

- The duration of tasks on a project is highly unpredictable. Task estimates on a traditional project have built-in contingency allowances.

- These allowances are often so engrained that task managers don't even realise that they are there

- These allowances are not padding – they are needed when uncertain events happen

- Managers much prefer certainty, and see overrunning a task estimate as 'bad'

- Task-level allowances for uncertainty are extremely wasteful

CCPM Approach:

- The allowance for uncertainty is made explicit in the schedule as a buffer

- Buffers are project-level allowances of time and money. The size of the project buffers can be significantly smaller than the sum of the invisible buffers built into tasks on a traditional project. This is why CCPM schedules can be shorter, and budgets lower

- Buffers are expected to be used. There is no negative connotation to using buffers

- The key project control measure is buffer management, where the amount of buffer used is compared to project progress. A simple graphical tool (such as a fever chart) provides a clear early warning so that issues can be resolved early and easily. This is one reason why CCPM projects are more reliable

- Buffer management is used to manage both individual projects and multi-project portfolios

CCPM Principle 2: Focus

The issues addressed:

- Members of staff on projects feel pressure to multi-task, or more precisely to switch between tasks before the first one is finished. This task switching (also known as 'bad-multitasking') is very inefficient, and is the main reason that tasks and projects take longer than they need to

- Pressure to show some progress, and to start as soon as possible, even though you may lack of all the parts/ information/resources to complete a task, increases task-switching. People tend to like to keep (or look) busy, so start something else that may not be ready to start, as-and-when their task hits a problem

- Due to the hidden safety in each task estimate, people intuitively feel that they have more than enough time for each task, so to keep busy they might start two tasks at once, believing that they are helping the project

- The lack of clarity of priority, especially for teams who work across several tasks or projects, can mean that project managers who shout loudest get priority. This leads to resources trying to keep several people happy, and to try and progress several tasks at once

- Progress reporting is infrequent and often subjective, with many project managers only seeing the overall picture monthly (and then the data tends to be at least a week old before being issued). This makes control actions rather slow and cumbersome.

CCPM Approach:

- Minimise multitasking and task switching
- Once a resource starts on a task, they work 100% on this task until it can be passed to the next stage, or there is a natural pause
- The project is run like a relay race – with tasks being the baton. When the baton arrives, you start and go as fast as you can until you pass it to the next person.
- CCPM schedules task starts as late as possible, whilst allowing sufficient buffer safety to complete the whole project on time
- Buffer management provides a clear priority signal to all project staff as to which task to do next – this helps avoid chaos caused by prioritisation based on 'who shouts loudest'
- Task durations are estimates, NOT commitments/promises. If problems arise, then a task will draw down from the buffer. If a task finishes early, then the successor task starts as soon as it can
- There is a strong management focus on 'full-kitting', which ensures that everything needed to complete a task is available before it starts. Projects often dedicate resources to just this role.

- Progress reporting is done frequently (typically at least once a week), using a very simple process. This ensures the buffer management process and fever charts are up-to-date, and any corrective actions can be rapidly implemented. Problems are identified and solved when they are small

CCPM Principle 3: Pipeline

The issues addressed:

- Many organisations run too many projects in parallel, they believe "the more you start, the more you finish". Counterintuitive though it might be, in order to maximize project completions, and minimize durations, you need to limit the number of projects underway

- This is a particular issue because of the key resources that are shared across the projects. Capex projects typically have less of an issue with shared resources than other types of projects, because most of the project team members are dedicated 100% to a project. Even so, key resources such as senior managers who oversee and give key approvals, are shared across projects. Also companies usually have a limited pool of trusted project leaders, which becomes over-stretched if the organisation takes on too many projects

- Having too many projects underway at once leads to bad multitasking, which in turn adds delay in inefficiency

- Trying to synchronise and schedule all resources across all projects is extremely complex, especially given the variability to actual task durations. In practice company-wide scheduling and resource management does not happen

CCPM Approach:

- CCPM includes a structured process to optimise the number of in-progress projects. It manages the portfolio based solely on the availability of your key resources. This ensures that this element of your business is not over- or under-loaded

- Projects that could be started are held back until the right time, when the new project can be executed without delay

- You will have only one or two resource types that are 'key' – i.e. their available capacity is a limiting factor on the whole organisation's capacity

- Accept that many resources will have some unused capacity. This is not waste, it allows you to focus on the one or two resource types where capacity is limited. A system where all resources operate at close to 100% capacity is extremely unstable and almost unmanageable.

- Projects should be sequenced based on the availability of the key resources; resources who you allow to focus on one project at a time. A strategic resource buffer of time is used to minimise the impact of delay in the key resource's work on subsequent projects

- "Finish a project before you start a new project" philosophy

Many CCPM implementations reduce the pipeline as step 1 – immediately stopping at least 30% of ongoing projects. This counter-intuitive move usually has a dramatic impact – fewer projects being worked on means a significant increase in the rate of projects being completed. One major Japanese healthcare electronics business stopped more than 90% of internal development projects when they implemented CCPM, which resulted in a 400% increase in the rate of project completions within a few months.

The Bottom Line

CCPM may seem trivial, just a collection of simple good practices, that 'we do informally'. Don't kid yourself. There are several features that are radically different from today's common practice.

Even for those practices that are used by today's better project managers, there is still benefit to be found in formalising them through the use of CCPM. Airline pilots and hospital staff use written checklists of 'simple good practices'. Despite their expertise and the number of times they have repeated a given process, it can happen that they overlook something when it matters most.[13] Project managers are no different, and CCPM formally embeds good practice into each project, and into the training of project managers. This drives consistency and quality.

CCPM works! Projects are delivered on-time in less time. They cost less through reducing waste, and they improve

---

[13] This topic is covered in depth in Atul Gawande's excellent book *"The Checklist Manifesto"*

quality through reducing errors due to multi-tasking and changes.

Businesses adopting CCPM have reported reductions in project durations of 25-50%. They have claimed that the same resources have been able to do more projects, at much lower stress levels. These improvements by themselves have led to increased profitability for both project clients, and the project supply chain.

CCPM is not just another way to manage projects, it is a significantly better way. Look back at the companies listed on page 27. Before implementing CCPM, they believed that they were good project managers, and many of them didn't think they had a major issue with projects. They still made rapid, sustainable, step-change improvements through implementing CCPM.

So why is CCPM not used more in CapEx Projects?

Almost all the success in using CCPM has been on projects where most of the project team works for a single organisation, and it has been relatively easy to establish a collaborative project team and implement shared time and cost buffers. It is also easier to control multitasking by shared resources who traditionally have come under pressure to support too many different projects at the same time.

But when the majority of the work is carried out by contractors, as it is with capex projects, things are not quite as easy. In order to exploit CCPM, the project team needs to be formed using a commercial process that encourages and rewards collaboration across the whole project team, and removes any incentives for contractors and suppliers to focus more on their own success than that of the project. This means rethinking the traditional contracting approaches used on capex projects.

You can work around this problem by just paying contractors a daily rate for the amount of time they work on the project. Many of the successful uses of CCPM on capex projects have taken this approach, with the client having a hands-on role in implementing CCPM, and managing resources.

But what if you want to use the professional expertise of contractors and suppliers, and to get more from them than just bodies to follow your instructions?

The next chapter is about how this can be achieved.

3

Collaborative Contracting & Project Alliances

"It is amazing what you can accomplish if you do not care who gets the credit."

Harry S Truman

Collaborative Contracting in practice:

- A collaborative contracting arrangement between the US Government's Department of Energy and the joint-venture Kaiser-Hill, *saved $30 billion, and completed a project 65 years early*[14]

- The UK Oil and Gas industries CRINE initiative in the 1990's, significantly improved the performance of major projects, *reducing capex costs by up to 30%, and cutting durations by up to half*[15]

- BP took the collaborative contracting approach used by CRINE to Australia. During the 2000's it became one of the

---

[14] Vitasek & Manrodt (2012)
[15] CRINE (1994)

Australian public sector's main contracting methods, between 2003 and 2008, it was used to deliver *over $26 billion worth of infrastructure projects*[16]

- In 1995 the chemical company Rohm and Haas had an important project in the UK. Their plant had never had a project of this size compete on-time in the past, nor on budget. Working collaboratively with two supply partners, the project was completed on-time and under-budget, *saving an estimated $5M*[17]

- In the USA, the Construction Industry Institute reviewed hundreds of construction projects in the US in the early 1990's. Those using collaborative contracting approaches were on average *10% cheaper and 20% faster* than less collaborative approaches[18]

Most capex projects have a preference for fixed-price contracting. It is seen as the best way to manage risk and uncertainty, and to remove the fear of being exploited by unscrupulous suppliers and contractors who, once selected and incorporated into the project, charge the client what they can get away with.

If you want to take advantage of CCPM, which we believe all projects should do, then fixed prices are not the right way to go.

Even if you do not use CCPM on your projects, then fixed prices are still not the best commercial approach to use for

---

[16] Government of Australia (2011)
[17] Personal experience of Ian who was involved in the project
[18] CII (1996)

The Executive Guide to
Breakthrough Project Management

the more complex contracts in place on capex projects[19]. The main reason for using fixed prices is to reduce risk in the project outcomes – a laudable aim. The irony is that they actually have the opposite effect. The cost and schedule uncertainty increases, and the whole investment and business case is potentially undermined. This is on top of increasing the total cost.

In this section we will introduce the *Project Alliance*, a 'collaborative', or 'relational', contracting approach, that can help to overcome many of the disadvantages with the more common commercial approaches used on capex projects. Project Alliances have been used successfully in a wide range of projects, bringing several advantages over the more traditional contracting approaches.

But before talking about Project Alliancing, we should explain how fixing prices can cause problems.

The Issues with Fixed Price #1: Risk pooling

One of the major issues in using fixed-price contracts was highlighted in the previous chapter; the idea of risk pooling.

Mathematically, the best place to account for the inherent uncertainty in a project is at the project-level. You will need a smaller allowance overall if you do not expect the

---

[19] *We are not against fixed price contracting per-se. When buying well-defined, simpler services and materials they make a lot of sense. It is on contracts where there is lots of uncertainty and complexity that we question their suitability.*

individual work package contractors to cover the cost of uncertainty and variation.

Fixed prices require contractors to include for all the variability that might arise in the work <u>they</u> are responsible for. Some contractors will need this allowance, others won't, and overall there will be more contingency included in the contracts than is actually required across the whole project.

This is not a trivial amount. In the previous chapter we described how most projects need less than 75% of the time that is allocated when each contractor manages uncertainty themselves.

On projects, time is money, so if 40% of the project cost is resources, the reduction in cost due to CCPM alone would be 10%.

If we add in the other cost contingencies that contractors need to allow in giving fixed-price bids, there could easily be a further 8-15% cost reduction available, purely from pooling the allowance for uncertainty at the project level, and not with each individual task or contract.

This means that contractors could be incorporating a premium of 18-25% for the convenience of having fixed cost and time commitments from work package owners.

In a simulation that we ran, without making unreasonable assumptions, we demonstrated that a project could be paying a 40% cost premium, just by insisting on fixed prices rather than reimbursing actual costs. You can see a short presentation of this example on YouTube (https://www.youtube.com/watch?v=lO0jUyhrOi4).

Note on project contingency:

Many projects already include some cost contingency in the project budget. These allowances are typically used for larger risks that have been identified during the planning stage. An allowance that is less than the sum of the cost to resolve each individual risk event is included in the budget. The project manager can then allocate this budget to those tasks that require it.

The principles we propose use exactly the same idea, but extends the scope.

A traditional contingency protects against larger events, with a low individual probability. We suggest that it makes sense for you also include for the large number of other variations that exist on real projects; events that are typically off-loaded to the contractors to manage, and are incorporated into fixed-prices. As well as a cost allowance for the resource buffer used by CCPM, the cost buffer might cover changes to prices in purchased materials and subcontracts, salaries, unexplained damage, and work errors/rework.

What if the client uses a single contractor?

If you are a project client, who contracts with a single contractor, you may feel that you avoid the premium of different contractors including for their own time and cost uncertainty. Unless your main contractor manages in a similar way to Breakthrough PM this is not the case.

It is much more likely that the main/prime contractor manages the project conventionally, and pushes the management of uncertainty down through the project's Work Breakdown

Structure,[20] either to their own staff, or the sub-contractors they use. This is commonly referred to as 'risk offloading', where suppliers lower down the supply chain are expected to manage a range of risks and uncertainties.

The coordination between the various sub-contractors requires significant effort, adding cost and time to the main contractor's estimate, and in turn to the prices they quote.

The Breakthrough PM approach is just as applicable to a single main contractor as it is to a project client who manages their own project. This topic is discussed further in Chapter 5 - Implementation.

The Issues with Fixed Price #2: Changes and Claims

It is a generally accepted truism in the contracting world that you "bid to win, and make money on the variations."[21]

Fixed price bidding encourages this behaviour, especially when it is combined with so-called 'traditional contracting', where the client engages an architect or design consultant to develop a design against which they can obtain fixed-price bids from competing contractors. This approach is also known as 'design-bid-build'.

---

[20] Work Breakdown Structure, or WBS, is a hierarchical representation of all the different packages of work on a project.
[21] This is not only an issue in capex projects and construction contracting. Authors Brown, Potoski & Van Slyke highlight this issue in section 6 of "*Complex Contracting: Government Purchasing in the Wake of the US Coast Guard's Deepwater Program*" (Cambridge University Press, 2013). The authors also discuss the benefits of relational contracting for complex contracts.

The tender documents for fixed price bids are usually detailed and complex. They will often have errors and omissions, especially since they were not prepared with any meaningful input from up-to-date construction experts (i.e. contractors). A contractor who wants to win the business, is discouraged from telling the client about the 'errors' during the tender stage, because the client would simply correct the errors, for all bidders. The bidder who spends the time reviewing the proposal and spots such errors thus gets penalised in the selection process, because any less-capable competitor would be given the information for free.

What if the bidder listed all the errors in their bid, and gave a quotation that included overcoming them? Firstly, in general, design consultants and architects don't like contractors drawing attention to their errors, so there is great risk that they are rejected for some kind of 'technical reason.'[22] Even if a bid passes the technical appraisal, there is always the risk that it produces a higher price. The Managing Director of a UK construction company told us:

> "Our price was about £32 million, and our submission included a list of reasons for this price, correcting much of the error-filled specification (many requirements conflicted with each other, and some were actually impossible to achieve!). The winning bidder's price was £14 million – less than half our price. The final outturn

---

[22] In 1993, when working in a global chemical company, Ian first heard the great term "*malicious obedience*" from a small UK contractor, who used it to describe their behaviour as a subbie! They knew much better ways than were specified in the tender, but they kept quiet and did what we asked. He had learned the hard way that well-qualified engineers really didn't like a subbie telling them what to do!

cost for the work was almost £45 million. Three times more than the 'fixed price' bid, and 34% more than our fixed price bid that the client deemed 'Way over the top'!"

Administering variations on a contract can employ almost as many people as managing the work! This all adds to the cost, and significantly extends the completion time of the project. It is not unknown for projects to spend several years after the physical completion, agreeing the final account.

The Issues with Fixed Price #3: It Makes the Project Longer

The typical bidding process outlined in the previous section wastes valuable project time.

Typical bidding documents for a fixed-price contract (whether 'traditional' design-bid-build, or design-and-build) tend to be measured in boxes rather than pages. For younger readers who have only seen electronic tenders, I guess we should talk of gigabytes, rather than kilobytes.

Since the design has been developed without the input of up-to-date construction experts, most projects expect the chosen contractor to review the design, and suggest changes. This results in duplicated effort and wasted time. The process is even built into the schedule as 'value engineering' or 're-engineering'.[23]   What this means in

---

[23] In the UK "re-engineering" is the term often used by contractors to mean the process to find what can be changed, and delivered at lower cost, in order to make more profit on the contract. Some changes would need client approval, whilst many will happen

practice is spending time doing the design over again to get it right.

This can be overcome to some extent by including a construction consultant in the design process, but this adds to the cost since you will get 'free' construction expertise with your chosen contractor. Another potential drawback is that it could reduce buy-in to the design from the actual contractors.

This means that the pre-contract design process is more focused on producing something that can be bid against, rather than finding the best solution to meet the client's overall requirements. Design is done twice, and once the contractor is selected, you have commercial tension arising from the contract, which can get in the way of full and open sharing of ideas. Overall, it wastes time, costs more, and tends to produce a sub-optimum design.

The second time-waster is the selection process itself. All the detail in the tender has to be read by the bidders, and then used to prepare their bids. Contractors rarely prepare bids using only their own estimating database; they get sub-contractors to estimate for them, as they issue tenders to them in turn. The sub-contractors then solicit further bids from manufacturers and other specialist sub-contractors. This all means that it can take months to produce a fixed price bid.

Once the bids are received they then need reading (and most fixed-price bids are not short documents),

---

unbeknown to the client. Most are perfectly acceptable alternatives, but on occasion they have been known to cut-corners, and not comply with the specification.

clarification meetings need to be held, shortlists must be prepared, and negotiations need to take place... All the while, the weeks and months pass by and time ticks away.

And time costs money.

The core project team still has to be paid. The longer the project duration, the lower the ROI. And even worse, if your new asset is needed to take advantage of a market opportunity, you lose months of income and gross margin.

Alternatively, you could continue with the detailed design process during the selection of the contractor. This would mean that you would have to continually re-issue the design and also allow the contractors to update their proposals during the selection process. This all adds time too.

And all this is for what? To choose a contractor whose bid price you know will not be the final project price![24]

What's more, all this wasted time has also cost the client in paying for the people who prepare and analyse the bid packages. This is not only the direct consultants they employ, but also those they pay indirectly. Preparing fixed-price bids takes a significant amount of time and

---

[24] Some clients take a very aggressive attitude to change once the contract is signed – "No!" is their stock answer. But think about it. Do you have contractors who are rich idiots, or are a charity whose role is to give money to the client, or will the bidders include a contingency to accommodate changes and hassle? If a contractor really has given you a low price, then they shouldn't be able to afford to accommodate change that incurs cost. We are aware of situations where contractors are rewarded for accepting changes at "no charge", by being allowed to "win" other fixed-price work to compensate them. Whatever way, the client pays!!

resource across the industry. The corollary of clients getting five competitive quotes is that contractors prepare five bids for only one to win – 80% of bids deliver no income. All this bidding takes time and costs money. This is cost that is included in contractor and supplier overheads, and ultimately recovered from clients!

The Issues with Fixed Price #4: It Reduces the Project Quality

Although issue #3 above was about the wasted time and money involved in fixed-price bidding, there is a further, potentially more serious consequence that should be noted. The quality is lower.

If the initial design (used for selection purposes) is done with low levels of up-to-date operational expertise, there is a significant chance that it contains errors and is more difficult to build. All the rework involved once the contractors are selected introduces an increased chance of errors if the consequences of these changes are not very carefully managed and communicated.

The Issues with Fixed Price #5: It inhibits Collaboration

Most of us understand intuitively how collaborative teams produce better results than lone individuals, no matter how talented and capable that lone individual. As Jim Collins said in his best-selling book *Good to Great*, if you get the right people 'on the bus', then you have a much better chance of success.

Best-selling author Patrick Lencioni highlights five key problems that can prevent a team from performing:[25]

1. Absence of trust
2. Fear of conflict (and the existence of artificial harmony)
3. Lack of commitment (to agreed actions)
4. Avoidance of accountability (not calling other team members to account and putting up with members doing their own thing)
5. Inattention to results (without common team goals, each team member has their own independent targets)

The way that contractors on capex projects are engaged encourages many – if not all – of these dysfunctions. This in turn reduces the probability of success.

Contracts Get in the Way

We are not going to address all of Lencioni's five dysfunctions here. We will focus purely on how the selection and contracting process can act as an obstacle to true collaboration[26] between team members. This relates to what Lencioni calls 'Inattention to Results' – the 5th

---

[25] From "The Five Dysfunctions of a Team". A summary can be found on Lencioni's website -
http://www.tablegroup.com/books/dysfunctions
[26] We say "true collaboration" to show that we do not mean lip-service, or interpersonal pleasantries. A collaborative team is one that has none of Lencioni's dysfunctions, where team members work together to achieve the common goal and their own local goals are subordinate to the overall project goal.

dysfunction - when team members focus on their own individual results, rather than the shared goal of the team.

Most capex projects have this dysfunction.[27] It happens whenever there is a way that a contractor can have a successful (profitable) contract, even if the overall project fails. It also happens when one contractor makes profit, whilst another suffers a loss, or when the client has a successful project, but the contractors suffer financially.

This dysfunction fuels the others – how can you develop deep trust between members when they are measured and rewarded differently? Lip-service is paid to common goals and non-contractual 'project charters' because they are not backed by money. Why raise a potential issue or suggest a simple improvement, when you can make more money by allowing the problem to develop, and receive a variation order or extension of time?

Most capex projects prefer to procure the services of contractors with a fixed price contract. The most common alternative is some form of reimbursement for the work delivered or time expended, based on pre-agreed unit prices. The preference for fixed prices is particularly strong with those contractors involved in the later stages; those who build the project and supply the major items.

---

[27] Of course it is not only on capex projects, where different team members have different objectives and measures. It happens within single organisations too, where senior managers mistakenly believe that you can manage a complex organisation with a series of independent functional objectives and measures.

These payment mechanisms have their place on projects, but they also cause serious damage. You can pay too much, and they can delay the project.

Fixed price (lump sum) contracting:

- Adds more time to the project schedule
- Usually increases the project cost
- Discourages collaboration amongst the project team, resulting in a poorer design and worse plan
- Prevents the use of CCPM

There are ways of working round these problems, for example using simple reimbursable contracts with the key project service providers and sub-contractors, and managing in a hands-on way. This is often an easy and quick fix for a project that wants to benefit from CCPM.[28]

However, this approach requires a significant amount of in-house resource from the project client, and even those who have used it would prefer to engage contractors for their expertise, rather than just as a supplier of feet on the ground. It also disincentivises contractors to contribute ideas, especially if they would reduce the amount of time they are paid for.

---

[28] The authors have spoken to people involved in a hospital building in the USA, and also large SAP implementation projects that very successfully used this approach.

Collaborative Contract Teams

Back in the early 1990's the US Construction Industries Institute carried out an extensive survey comparing the results of traditionally managed projects with what they called 'project partnerships', based on a single integrated project team (CII, 1996). Their results showed that projects using a collaborative project team delivered projects with:

- 10% lower total cost
- 25% higher supplier profitability
- 20% shorter durations, and fewer changes
- Much higher safety performance (e.g. 60x better lost-time accident rate)
- 50% reduction in rework
- 83% reduction in claims
- 30% increase in job satisfaction

Around the same time, others came to the same conclusion – an integrated team, working collaboratively, produces better results. (See Latham, 1994; Egan, 1998; CRINE, 1994; and the Lean Construction Institute, formed in 1997 (www.leanconstruction.org)).

Advance 20 years, and collaborative contracting on capex projects has still not become the norm, and organisations are still struggling with the dilemma that although the idea seems a good one, it's difficult to see how to make it work. Many attempts at project collaboration have failed to deliver significant benefits, and it is often seen as difficult and costly to set up, not to mention long-term in nature.

These are especially important considerations since businesses need short-term results.

Rather than diminishing our belief in the potential of project collaboration, this has underlined how important it is to do it in the right way.

Project Alliancing & Integrated Project Delivery Teams

Project Alliancing is a collaborative form of contracting that aligns the financial interests of the contractor(s) who work on the project, with those of the project client. It is an example of what is also known as a 'relational contract'.

Project Alliancing as a contracting approach, came to prominence in the UK, as part of the Oil and Gas industry's CRINE initiative during the early 1990's. Alliancing was a key part of the improvement programme that resulted in projects being delivered in half the time, and for two thirds of the cost of a more traditional project. (CRINE, 1994)

Project Alliancing has many similarities to IPD (Integrated Project Delivery), and Project Partnering, terms that have been used by the Lean Construction movement and the CII (Construction Industry Institute) respectively.

Whilst alliancing, when implemented correctly, delivers significant improvement in project performance, it has not been widely adopted as we believe it should be.

In recent years, one of the largest users of Project Alliances is the public sectors in Australia, where Project Alliancing has become a mainstream approach, especially for complex projects.

Project Alliancing started to be used in Australia on Oil and Gas capex projects in the 1990's. By the late 1990's, its use had spread to the public sector. During the 2000's Project Alliances became a mainstream strategy used to deliver major infrastructure projects, both in Australia and New Zealand. Between 2004 and 2009 infrastructure projects with a value of over USD 26 billion, (almost 30% of total infrastructure investment) were delivered in Australia using Project Alliancing (State of Victoria, 2009)[29].

A Project Alliance is formed by a client and one or more contractors/suppliers (referred to here as supply members). Each supply member should have a large enough role to influence the overall project outcome, meaning that in most circumstances, smaller suppliers would not be members of the alliance[30]. An alliance would typically have between two and seven members, including the client.

---

[29] This government audit also said that across these projects, there were no disputes that required external (and therefore costly) dispute resolution. In our view this is unprecedented, and would be a reason in itself to use alliancing – a significant saving in time and cost.

[30] An exception to this might be project management consultants, who although their package might have a low monetary value, their impact is significant.

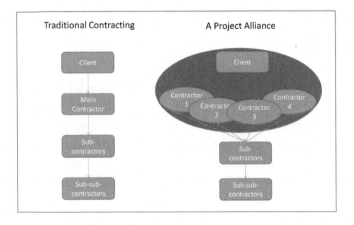

The 'client' in the Project Alliance does not have to be the organisation that will take ownership of the finished asset, though this is the most common situation. It can be used by a Design-and-Build or EPC[31] main contractor who is the sole supplier to the client, and who uses the principles of Project Alliancing with their supply chain.

The Critical Success Factors for an Alliance are[32]

1. An integrated team, with members selected based on competence and "best person for the role"
2. Collective sharing of risks and opportunities across the team

---

[31] EPC = Engineer – Procure – Construct
[32] This list of CSF's is based on a discussion with Jim Ross, founder of PCI Group, and one of the pioneers of the use of Project Alliances in Australia.

The Executive Guide to
Breakthrough Project Management

3. Contract terms that exclude the idea of "fault" and "blame", with minimal legal recourse for dispute resolution.

4. A payment scheme that fully reimburses variable costs, and aligns margin to the overall project success, rather than the success of individual work packages.

5. Unanimous, principle-based, decision making

How do Project Alliances work?

In this short Executive Guide, we will focus on one specific aspect of a Project Alliance, the part that is most different from traditional contract models – payment. This is the key mechanism used to align the commercial interests of all the Project Alliance members with the performance of the overall project.

The starting point is a 'target cost' for the whole project, which all the parties agree to. This agreement usually takes place during the selection and negotiation stages.

Defining and agreeing a total cost does take some time, and the details are beyond the scope of this Executive Guide. Because it is not as crucial as the price in a fixed-price contract, it does not take as much time or effort to agree. This is because the supply members are not taking as much risk on just this single figure, compared to the traditional fixed-price bid.

Once a total project cost, including contingency allowances, is agreed by all the parties, it is first nominally allocated across all the supply members, and then divided into two core elements: cost and fee.

- Cost means all monies that pass through the supply members, either to their supplier and sub-contractors, or their employees who work on the project
- Fee means the amount they expect to make as a contribution to corporate overhead and profit.

Each supply member will agree a nominal fee value based on their nominal role[33] in the project. The fee can then be further subdivided into:

- A *fixed fee*
- A *variable fee*, linked to the overall project performance (NOT the performance of the specific supply partner)

Whilst there is always a variable fee, fixed fees may or may not be included in the agreement. Each supply member will have a nominal variable fee which is the amount they will receive if the project goes as planned. The actual variable fee paid, usually on the completion of the project, can rise or fall depending upon the overall project success.

Actual Variable Fee = (Nominal Variable Fee) x (a factor linked to the project performance)

---

[33] We talk about their "nominal role" because once the alliance contract is in place, and the project is underway, who does what task can be varied based on skills, availability, cost, or whatever criteria the project management team wish to use. There is no financial incentive for any one alliance team member to increase their turnover, nor is there any penalty if it reduced.

The fees can be negotiated as part of the commercial selection process, or be pre-defined by the client. A Project Alliance in the mid-1990's that Ian worked on, used the published financial statements of the two supply partners as a basis to calculate the fixed and variable fees. The variable fee was proportional to the operating profit, and the fixed fee was in proportion to the corporate overheads.

We call this payment method CFV (Cost, Fixed, Variable), representing the three parts of the total payment).

Once the project is underway, each supply member is paid a regular payment (typically monthly) to cover:

1. Costs incurred, based on invoiced values from suppliers and sub-contractors, and employment costs for agreed 'direct' employees working on the project
2. An agreed proportion of the fixed fee. For example, if the project was planned to last 20 months, each month the fee might be 5% of the total fixed fee. Payment can also be linked to interim progress milestones, rather than the calendar.
3. On completion of the project each supply member receives the same multiple of the variable fee – typically between zero- and two-times the nominal fee value. Performance that achieves all the agreed project performance targets would receive the nominal fee (i.e. the multiple would be 1)

The principles behind the CFV payment method are:

- Once the project is underway, the only way for each supply member to make more profit is to ensure the whole project is successful, and thus increase the actual variable fee paid. There is no other mechanism for them to make more money. They can increase their turnover if they incur cost on behalf of the project, but they cannot increase their margin/profit

- Blame and fault are irrelevant. Just like in team sports, there is just one scoreboard for the team – we either all win or we all lose

- Unnecessary spending reduces the supply members' profitability, because one of the performance measures linked to the variable fee will be the total project cost

- Suppliers are not financially penalised for reducing the cost of their part of the project, in fact they benefit financially if cutting cost is in the project's interest

- Cost has to be true cost, with no hidden additional elements. Suppliers and sub-contractors' invoices should be net of any retrospective rebates. Staff salary costs should be based on real salary plus any salary-related charges such as pension, employer taxes, and benefits. Whilst it is beneficial to simplify the administration of this by agreeing to rounded up rates, it is important that there is no significant[34] overhead contribution hidden in the rate

---

[34] Significant means that any opportunity to gain margin by incurring additional cost should be significantly smaller than the variable fee lost due to over spending.

A Worked Example of Using CFV Pricing on a Project Alliance

Step 1.   The total project cost is agreed at $50M

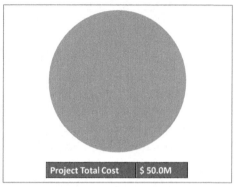

Step 2.   There are 4 supply members of this example Project Alliance (SM1-4 in the tables below). The $50M is nominally allocated between them, based usually on their particular fields of expertise. The allocation is 'nominal' because once the project is underway they do not have to deliver the scope in this allocation (though they usually will).

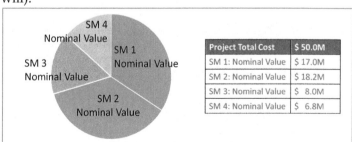

Step 3.   The nominal allocation is further split into cost and fee, still totalling $50M.

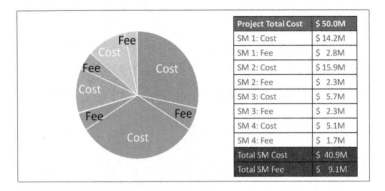

Project Total Cost	$ 50.0M
SM 1: Cost	$ 14.2M
SM 1: Fee	$ 2.8M
SM 2: Cost	$ 15.9M
SM 2: Fee	$ 2.3M
SM 3: Cost	$ 5.7M
SM 3: Fee	$ 2.3M
SM 4: Cost	$ 5.1M
SM 4: Fee	$ 1.7M
Total SM Cost	$ 40.9M
Total SM Fee	$ 9.1M

The fee may be negotiated with the client, or based on common principles across all supply members. Note the cost:fee ratio does not have to be the same across all the supply members.

The fee is further sub-divided into fixed fee and variable fee. For simplicity we haven't broken this out into individual supply members.

Step 4. This leaves a cost target (excluding fees) of $40.9M

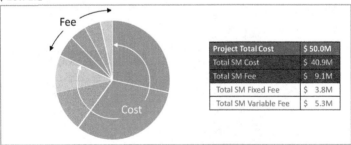

Project Total Cost	$ 50.0M
Total SM Cost	$ 40.9M
Total SM Fee	$ 9.1M
Total SM Fixed Fee	$ 3.8M
Total SM Variable Fee	$ 5.3M

In the table, the fee has been divided into fixed and variable elements. The split between fixed and variable fees shown is purely indicative. A typical range would be 30 - 100% variable fee, 0 - 70% fixed.

Each supply member will have their own fixed and variable fees defined, and the ratio between fixed and variable may be different between different supply members.

Step 5. The variable fee is then divided across a small number of project success criteria. Typically, 3-7 criteria are used, and might include; capital cost, time, quality/performance of the finished asset, operating cost, safety, or even the satisfaction of defined stakeholder groups.

In this example let us assume that the following three criteria are used; cost, time, and safety.

In this example, the client decides that the relative importance of these three criteria are:

- capital cost = 25%
- safety = 25%
- completion time = 50%

This gives the total variable fees shown in the table.

During the final stages of contract negotiation, the alliance team partners agree the performance levels that will trigger each

Project Total Cost	$ 50.0M
Total SM Cost	$ 40.9M
Total SM Fee	$ 9.1M
Total SM Fixed Fee	$ 3.8M
Total SM Variable Fee	$ 5.3M
Variable Fee - Cost	$ 1.33M
Variable Fee - Time	$ 2.65M
Variable Fee - Safety	$ 1.33M

variable fee payment. The principle is that a 'reasonable' performance will result in the nominal variable fee being paid. Better performance will result in a higher fee, worse performance a lower fee. Performance level commitments can be part of the competitive selection process.

In our simple example, an illustration of the variable fee to be paid is shown in the following graphs. In the contract there will be more detail describing how each of the criteria will be measured and payments calculated.

We have only shown the total variable fee. Each supply member of the alliance will have their variable fee divided in the same proportion across all the success criteria (i.e. 25:25:50 cost:time:safety in this example).

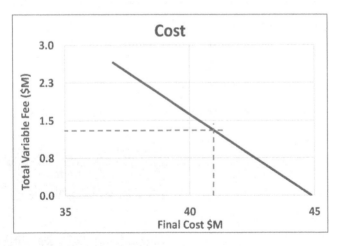

The red line shows that coming in on-budget ($40.9M) will pay the variable fee at risk of $1.33M.

The Project Alliance will negotiate a basis for sharing cost over- and under-runs, and whether there are any caps or changes in allocation. This graph is based on a split of 33% supply members, 67% client, meaning that a saving of $1M will increase the variable fee paid to the supply members by $330K, and the client will save $667K against the budget.

The completion date fee shown below works in the same way.

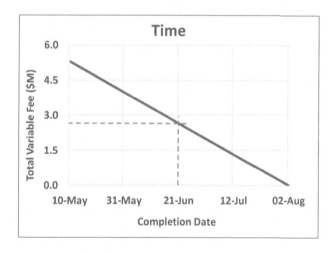

But how do you measure safety?

There are people who argue that 'safety is a given', and therefore shouldn't be part of a variable payment system. Whilst we agree that it is not optional, there are certainly a wide range of approaches to managing safety, and a range of results achieved. This example is based on three projects where safety has been included in the variable payment scheme.

On one of these projects, safety was the largest element of the variable payment scheme – that client believed that if safety really is the number one priority in their business, *"We should put our money where our mouth is, and expect our supply partners to do the same."*

That project developed a 'Safety Index' measure, which took account of the quality of the project's safety programme (acknowledging the efforts made), as well as the number of serious injuries (defined as injuries that result in more than 3 days lost time). Efforts made gained points, injuries lost them.

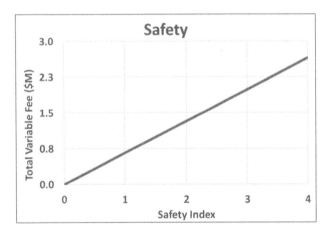

So for example a world-class site safety programme would gain 4 index-points, whilst an average programme would gain 2 points. Each serious injury would lose 2 index-points. Non-compliance with the agreed safety programme would also lose index-points.

The above example shows the basic structure and principles used to develop shared performance measures for a project alliance. There are many variations on this theme, and even relatively complex concerns can usually be accommodated. The key steps are

- Agree the important client outcomes from the project
- Decide how to link the variable elements of the fee to these outcomes
- Ensure there is no mechanism for supply members to gain reward by focusing on one outcome at the expense of others

Other elements of Project Alliances

Payment method is not the only difference when using Project Alliances - all that using a CFV payment scheme as described above does, is to remove commercial obstacles to collaboration.

We have described this in detail because it is, in our opinion, a pre-requisite to the success of Project Alliances.

Whilst there are examples of collaborative project teams that have used traditional contracting methods, the success rates are much more variable.

If a project goes well, and the contractors can make reasonable profits at the same time as collaborating with the client and other contractors, then traditional contracting can work well (though the project will probably be more expensive than it needs to be).

However, if things do not run smoothly, or a contractor discovers that they have underbid for the work, then their prime focus becomes money. If in order to make money they need to ditch the collaboration across the project team, then so be it! Contractors end up fighting for survival, and cashflow becomes their number one priority.

Aligned financial incentives alone are not sufficient to guarantee project success. It does however remove a major obstacle. The project manager will still need to actively manage the building of trust across the team and at senior management level between the companies involved.

The "Sunflower Model" shown below summarises the main factors necessary for a successful relational contract (like a Project Alliance).

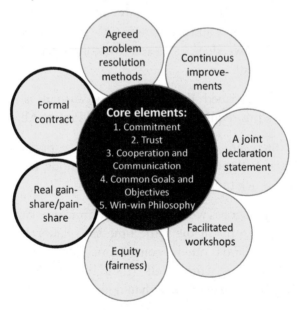

The Relational Contracting Sunflower Model
From Yeung, Chan & Chan, (2012)

The model is described in an article published in the International Journal of Project Management (Yeung, Chan and Chan, 2012), where over 50 separate studies and references on relational contracting were reviewed to try and identify a common definition of what relational contracting involved. The paper highlighted that whilst there was some variation, there was strong consistency across all 50 research papers. Almost all cases contained the core elements (at the centre of the flower), along with real share in gain/pain incorporated in the formal contract.

Whatever process you use to establish a collaborative project team, the team still has to exploit the collaborative team environment, by applying project good practices to reduce waste and speed up the process. Many of the value-adding practices that projects can use are enhanced under a Project Alliance, because the whole team is involved in them. This includes practices such as value analysis and engineering, risk management, information sharing and design models, and safety management.

As described in the previous chapter, CCPM has to have a collaborative team for it to work, and the idea of a shared buffer is incompatible with fixed prices. CCPM also helps build the project team by building collaboration into day-to-day activities.

Sharing the buffer status using the fever chart reinforces the idea that the whole team is 'in the same boat'. The

UK's ACTIVE[35] initiative from the 1990's used this analogy with the following cartoon, to highlight the idea behind project team collaboration.

What it does particularly well is to highlight the need for aligned hearts-and-minds, as well as aligned incentives. The two men in the dry end of the boat still have their traditional contracting mind-set of fault and blame. Under a Project Alliance, their interests are best served by helping

---

[35] ACTIVE = Achieving Competitiveness Through Innovation & Value Enhancement, was a government-industry initiative aimed at improving capital project performance in the UK chemical and process industries during the late 1990's. Ian was involved leading one of the pilot projects, and was a member of several ACTIVE working parties, involved in drawing up best practice guidance.

solve the problem. A Project Alliance should engender the Musketeers' spirit of "One for all, and all for one."[36]

The frequent reporting and progress updates that feature in CCPM also help to embed the collaborative spirit. Daily updating of progress, and short review meetings give ample opportunity for project leadership to lead from the front. The structure gives the perfect chance for project leaders to demonstrate that the traditional project behaviours of deflecting blame, and achieving the planned task completion date, have no relevance or importance on a CCPM project.

Whilst the idea of more frequent project reports and meetings might not sound like a 'good thing' to many of us, under a CCPM project they are very different. Since fault and blame is irrelevant, all the focus is on completion, helping keep meetings both short and of interest to all parties. They focus on "What is preventing us completing sooner?", and "What help do task teams need to overcome obstacles?". Many projects hold stand-up progress meetings, using visual displays or manual boards, making sure they last minutes rather than hours. Progress reporting, done directly by task owners on the CCPM software system, results in an immediate update to the fever chart, making it very easy to use up-to-the-minute information in making management decisions, and giving the earliest warning of potential problems.

---

[36] *A traditional saying used to confirm strong team camaraderie, made famous by Alexandre Dumas in his 1844 novel "The Three Musketeers"*

The cost and time buffers are also great collaboration enhancers. Because they are the team's shared protection from uncertainty, they provide a simple and visual indication of how 'we' are doing against 'our' targets.

In the next section we discuss a few other common value-adding methods used on capex projects, and show their compatibility with Breakthrough PM.

The Selection Process

Selecting Project Alliance members is also significantly different from the more usual project tendering approach.

When appliance members are chosen, we are much more interested in their competence, and the availability of good people to work on the project, than their (gu)estimate of how much the project will cost. Selecting a main contractor based on a fixed price is like selecting a footballer for a team based on their claims of how many goals they will score next season[37].

All too often we hear of a selection process being used that looks for the 'lowest-price technically-compliant' candidate, wherein the candidates are assessed to see if they meet a minimum standard, and the cheapest of those is chosen. We suggest it is far better to make balanced and informed decisions.

With football, more important than price (salary and transfer fees) are criteria such as the player's skill, experience, and how well they will fit into the rest of the

---

[37] Since we are British and Australian, we have written in UK-English, so our reference to "football" is what Americans call "soccer". We think the analogy works just as well with American Football.

team. Cost, of course is relevant, but this is not the main criterion. Interestingly, the characteristics of the footballer in this analogy are very similar to the characteristics that we hope to see in our project alliance team members – skill, experience, and team fit.

It is important to stress here that selecting a collaborative team member primarily on competence does not mean that the contracting process is not competitive.[38]

A good selection process:

- Is short (4-8 weeks is quite achievable)[39]
- Will go through several stages (long-list, medium-list, short-list)
- Assesses the underlying organisation, their senior management, and the people they propose for the project
- Focuses on the contractor's ability to manage both their specialist technical area, and also as a member of the wider project team. (Assessing recent past performance is a key part of the assessment)
- Will take place much earlier in the project than is typical today
- Uses a weighted criteria process to compare proposals

---

[38] Nor does it require that you work with the same contractor or team across all your projects. Whilst multi-project framework agreements do have a role, they are neither required for Breakthrough PM, nor are they necessarily a good approach to take. They are also beyond the scope of this book.
[39] This is to select, not to enter contract with, though you will have agreed the most important elements of the eventual contract during the selection process.

- Shares the client's project overall cost and time estimates with the bidders. At the short-list and final negotiation stage, the estimates will be shared, and an agreement made on the contractual cost and time targets.
- Uses fixed and variable fees as the main commercial criteria[40]

A typical selection process might involve:

1. Written submissions from a <u>long-list</u> of bidders from which;
2. A reduced number (<u>medium-list</u>) will be invited to a second stage, typically involving a presentation of their proposal and discussion with the selection team. Here they will be assessed based on criteria such as competence, their ideas for how to deliver the project, and their capability of the key team members, before you produce;
3. A <u>short list</u> (usually one or two) who move to the final stages which may involve visits to the bidder's offices and project locations, as well as meeting current clients.

---

[40] Since the fee value is small relative to the overall project cost, it is risky to weight the fees highly in the selection process. Capability is much more important.
You can include the cost target as part of the commercial assessment for simple projects where bidders have enough intuition and experience to quickly calculate a "ball park" cost target. Remember this is NOT a quotation; it is a target that they are willing to link part of their variable performance fee against. However, we do not recommend doing this where the bidders would need to spend significant time estimating the project cost.

One of us (Ian) used this process back in the mid-1990's when a Project Alliance partner was selected using this method in four weeks, for a $30M process industry project (2016 values). The alliance had three members; client, engineering consultant, and construction partner.[41] The RFP[42] was less than 4 sides of A4. There were 8 bids received, 5 companies were invited to present, and 2 were shortlisted. Dates for the presentations and site visits were set in the RFP, which allowed the entire process to take less than a month.

The Contract

We recommend that you start with a blank sheet of paper, ideally a flipchart.

In the room you have the senior responsible managers from all the alliance partners. At this stage you don't need a contract specialist.

The goal is to agree the core elements of the contract, and at the same time, start to build the team, and develop the trust that will be needed to make the project a success.

An independent facilitator is also a great idea, because it means that each attendee can focus on their role as a team member, rather than having to try to multitask and facilitate the session as well as contributing.

---

[41] The engineering consultant was already selected and had a contract to produce the front-end definition. The Project Alliance was formed once the construction partner was selected.
[42] RFP = Request For Proposal

There will be several outputs from the contract-development and team-building session(s). Here we will look just at the contract.

The team will discuss and agree the key elements across a range of topics including the overall objective and critical success factors, the payment mechanism, the fixed and variable fees, how performance is to be defined and measured, which risks are included in the alliance and which are retained by the client, how the overall project will be managed, and how any disputes will be handled. It is also important that, in the positive atmosphere often present at the time of project commencement, the team also discusses what will happen if things change or go wrong.

The flipcharts are then written up into a 'Heads of Agreement' document, written in plain language, and agreed by the senior representatives. It will usually take several iterations of this document before everyone is happy with it.

Once agreed, the Heads of Agreement is passed to a contract drafter to turn into a complete agreement. Their brief is to ensure the words and intent are embedded into a complete, legally sound, document. We also highly recommend that the contract writing follows the principles of 'simpler contracts', using plain language and avoiding legal jargon. This helps to ensure that the final document is a practical "user guide" for the Alliance, rather than a traditional, "put-it-into-the-bottom-drawer-and-hope we-never-need-to-read-it-again", contract.

This final contract can either be a bespoke contract, or based on a set of model contract forms. If a model form of contract is used, it is important that it is suited to a Project Alliance, and the CFV payment method[43]. We do not recommend that you try and modify a non-collaborative contract form.[44]

Typically, one of the alliance member's contract specialists will do the drafting, with the other members reviewing and commenting.

Everything is now in place.

The project team members now have a joint destiny on the project – either everybody wins or everybody loses.

This is the ideal environment to exploit CCPM to deliver the project on-time in less-time, and at lower cost with higher contractor margins.

The benefits from a collaborative project team aren't limited to using CCPM. Since the team is in place much earlier than usual and the members each have a stake in the success of the project, this helps a range of other project improvement techniques to produce great results.

---

[43] At the time of writing, most of the common model contract forms used in capex projects, including the NEC and FIDIC families, do not include a variant suitable for a Project Alliance.

[44] Ian did this in the early 2000's, when rank was pulled, and the IChemE Green Book terms were required to be used. It took at least three times longer, and was much harder to read and understand than a bespoke contract.

The next section looks at some of these techniques and how they fit with CCPM and Project Alliancing in more detail.

4 Other Project Management Methods

Breakthrough PM uses CCPM to schedule and manage project progress, and Project Alliancing to remove commercial obstacles to teamwork.

The collaborative project team that is established with this process is much better placed to exploit a wide range of project management tools and techniques than would otherwise be the case.

This section describes a few of the most commonly known project management methods that are highly compatible with Breakthrough PM.

We also mention a few of those we believe to be incompatible. This doesn't necessarily mean that the methods are bad – many are significantly better than nothing.

But in the same way that it's equally valid for a country to drive on the left or right hand side of the road, it has to be one or the other, or chaos will ensue. It is the same with the key pillars of Breakthrough

PM; CCPM and Project Alliancing – you can't implement them, whilst at the same time retaining your old practices.[45]

Highly Compatible

Method	Comments
Value analysis & engineering	You should get better results from the process, since the delivery team members are involved early, and there are no financial disincentives to highlighting potential improvements.
Collaboration technologies	Project databases and information sharing systems are important on large projects. Some traditionally-contracted projects have problems with sharing data and ensuring compatibility between members. Invariably standard contract forms lag behind advances in technology.

A Project Alliance should encounter no such resistance to sharing. There should be one source of 'the truth' and a single system that all project members use. All members should have access to a single information management system.

If there are debates, they will be about which system to use, and whether it really helps the project to be successful |

---

[45] You can of course pilot Breakthrough PM on a single project, providing it is isolated from the main organisation during the pilot. If you have many key resources that are shared across projects, piloting on a single project is more difficult.

	and adds value, rather than arguments over 'who pays?'
3D Design models & BIM	These are similar to the above. Their project-wide use should be made much easier across an alliance project. With the range of different systems available, the ability of alliance members to use technology and to integrate into a single system can sometimes form part of the selection process, even if only as a differentiator between two similarly capable bidders.
Lean Construction	Lean Construction and the use of methods such as LastPlanner®[46] are compatible with Breakthrough PM, in particular where Lean Construction methods are used to reduce the critical chain, and to synchronise short-term work schedules. CCPM does not seek to schedule in detail, and so collaborative short-term planning methods, such as LastPlanner®, can easily be used on a CCPM-managed project. In addition, the establishment of collaborative contracts with the main contractors removes any commercial tension that may inhibit day-to-day planning and coordinating between different contractors – they are 'all in it together', and anything that improves

---

[46] LastPlanner® is a methodology licenced by the Lean Construction Institute (http://www.leanconstruction.org/training/the-last-planner/)

	day-to-day site synchronisation, helps the alliance to achieve the common goal.[47]
Agile & Kanban approaches to PM	CCPM does not try and schedule and manage detailed tasks. Task managers are allowed to oversee their own local workload as they wish, so long as they follow the CCPM principles of focused work, frequent and honest reporting, and follow the CCPM priorities across tasks. In several environments, project teams have used other scheduling and execution techniques such as Visual Project Management and task-level Kanban boards.[48] These have successfully blended using the Agile/Kanban method to manage daily detailed tasks, whilst using CCPM for the overall project management. These approaches have been particularly successful in IT, and also on large one-off manufacturing and maintenance projects.
PDRI – Project Definition Rating Index	PDRI is a method that was initially used in the process industries for evaluating the quality of the early stage project definition activities. It helps to prevent starting construction too early, resulting in aborted work and high execution risk.

---

[47] Unlike one "lean construction" project that Ian was involved in. Through the application of the lean toolkit on the site, one of the main sub-contractors finished their work 6 weeks early – a reduction of almost 20%. However, because the overall project was not integrated into this process, this brought no real benefit to either the main contractor or the client. They stuck to the original plan.

[48] For a quick overview of visual project management ideas see Mark Woeppel's eBook, "Visualising Projects", available from http://pinnacle-strategies.com/go/visualizing-projects/

	There is a strong correlation between the PDRI score before starting construction, and a project's cost and time performance. PDRI was developed by the CII in the USA and versions exist for different types of capex project. Since a collaborative project team has incentives aligned to the client's business objectives, they can use techniques such as PDRI to improve the project chances of success. Also, Project Alliance members who were not involved in the front-end definition can use it to assess the quality of a project's plan and design. The early involvement of execution-stage contractors helps to improve the quality of the project plan, and thus the PDRI score. No one has a vested interest in starting early or increasing change and design rework. CCPM also uses the concept of "full-kitting", which generalises this idea, and reduces the inefficiency when tasks are started too soon.
Risk Management	Project Risk Management is a well-established methodology for identifying, reducing, and planning for, the wide range of risks that projects are exposed to. It is best done with a wide-ranging cross-functional team. Construction-stage risks are best identified early and with the input of experienced construction specialists. A Breakthrough PM team provides just that environment.
Public-Private Partnerships (PPP)	Also known as P3, 3P and PFI (Private Finance Initiative). These are methods used in the public sector to fund infrastructure using private finance to supplement, or replace, public funds.

	The public sector, or user, pays a charge to use the asset over an extended period of time.
	PPP is NOT a method of project management and delivery; it is a method of funding the project.
	We see no reason why a PPP project cannot use the methods of Breakthrough PM, to deliver the project faster and for lower cost.
	There is a question over whether the PPP approach to the funding of infrastructure investments provides long-term value for money to the sponsoring public entity. We won't comment on that here, but we do believe that a PPP project that uses Breakthrough PM could improve ROI and reduce risk for the investors, whilst reducing subsidy, risk, and delay to the public sector.

Not so Compatible

Method	Comments
Non-CCPM planning software	This is a round peg in a square hole! You _can_ manually schedule tasks to be the buffers, but there is no calculation of the critical chain, no automatic integration of buffers, nor reporting automation. Some software claiming to incorporate CCPM takes this approach, and we would not recommend its use. If the software does not allow simple frequent reporting by task managers[49], or support buffer management with simple and clear red/yellow/green task status indicators, we suggest you find another system. It is possible to create a traditional view of a CCPM programme, for cases where a client insists that this is produced, but this is unnecessary extra work, and can trigger unhelpful interventions from influential stakeholders. If the client insists, just do it, but make sure it is only used to communicate to the client and not used by the project team.
Earned Value (EVM)	This is fundamentally inconsistent with CCPM. In our opinion, CCPM provides a much more robust method of progress management and early-warning than

[49] Under CCPM, task managers provide a regular update of the expected task completion date or time outstanding, and _not_ of the % complete. This will usually be at least once per week, maybe even daily on shorter projects. This is why it is crucial that this is an easy thing to do, otherwise it will only become a bureaucratic waste of time.

	EVM. It is possible to produce EVM progress data from a CCPM schedule, for cases where a client insists that this is produced, but this is unnecessary extra work, and can trigger unhelpful interventions from influential stakeholders. Bear in mind that if you decide to do this, you will need to adapt the Critical Chain schedule in order to account correctly for the buffers[50] and create an EVM baseline.
Interim Milestones	These can be accommodated in the plan, but they tend to distract more than they help. Most milestones are clients' methods of reducing risk and improving performance. Using CCPM, Breakthrough PM provides a better way of reducing the risk. Remind the client that when using CCPM, the focus is on project completion being on or before schedule, and that they will have constant visibility of this through the project buffer. As with other methodologies in this section, if a client insists, interim milestones can of course be integrated into the schedule. Very important milestones, for example those attached to an interim payment, or those linked to other projects, should be protected with their own buffer in the CCPM schedule. If you are managing very large projects, we recommend that you split the project

---

[50] A whitepaper discusses a method to do this on the book website.

	into smaller sub-projects, and manage these as a programme of linked projects. Overseeing progress with a portfolio fever chart, rather than blunt single date milestones should help you here.
	This approach gives the overall programme manager early warning of any problem projects, which is something simple milestones don't give you.
Fixed Price Contracting	Two core ideas that underpin the Project Alliance are 'no blame' and aligned incentives, which are incompatible with the use of fixed prices. A fixed price is used to allocate risk and responsibility to the parties to the contract. If something goes wrong, the contract requires one party to take the blame.
	Another consequence of using fixed prices is that the project needs to do a fairly detailed design in order to select a contractor and contract to a fixed price. This design is therefore carried out without any input of the people who are the experts in execution and construction.
	The Project Alliance in contrast, looks to choose the alliance members as early as possible, based on primarily on competence rather than price.
Cost-plus Contracts	Technically the CFV fee method that we recommended in section 3 is 'cost-plus'.
	What we are referring to here is the cost-plus-a-percentage-of-the cost method.
	This type of contract is incompatible with Breakthrough PM because it rewards and penalises the wrong behaviours. If a contractor comes up with an idea that

can achieve the project goals, but uses less of their resources, then they reduce their own financial income. In a similar vein, if they want to increase their own profitability, the only route this contract type leaves open is to increase the billable work they do on the project and/or the price they pay for purchases and sub-contracts.

The other way it harms the project team is that it encourages the contractor project alliance members to inflate their own role on the project. For example, it might be in the client's interest if a different member carried out a specific task, but a member who points this out would harm themselves financially under a simple cost-plus payment regime.

5 Implementation

"[If most organisations implemented what is known today]...

... the rare firms that are able to consistently translate knowledge into action would not enjoy the substantial competitive advantages that they do."

Jeffery Pfeffer & Robert Sutton, Stanford University, in their best-seller "*The Knowing-Doing Gap: How Smart Companies Turn Knowledge into Action*"

Warning: Change Needs Managing

The above quotation is over 15 years old, and is from a book published just after Oil and Gas companies learned how successful Project Alliancing could be, and after one of the UK's largest construction companies proved that CCPM could deliver significant reductions in the duration of construction projects.

In *The Knowing-Doing Gap*, Pfeffer and Sutton, show that the failure of organisations to exploit and embed the competitive advantage that came from the knowledge they had acquired, is sadly, much more common than we might believe.

Their key conclusion was that good ideas don't embed themselves just by being good. Companies need to actively manage the process, and few do it well.

In almost all organisations, implementing the ideas in this book, whilst simple to understand in concept, will not be easy. It will need more than an email from the CEO, or sending a few project managers on training courses. Without the full and active support from the executive leadership team, it is unlikely even to get off the ground.

It will involve changing well-engrained practices and challenging widely held assumptions about the so-called 'best ways' to do things. Whilst the practicalities of managing this change are beyond the scope of this short book, we did not want to leave you with the feeling that just because the concepts may be simple to understand, then it follows that they will be easy to implement.

However, we don't want to put you off either! They can be implemented quickly, and deliver results quickly. But it will take time and planning to embed sustainable change into any organisation.

Take for example when I (Ian) first worked on a project alliance in the 1990's. On that project no one had ever worked on an alliance before. But with a strong desire to do so, and some part-time advice and facilitation, it was not difficult. The senior management of all three alliance members was fully supportive, and the project leadership team was allowed to get on and implement the alliance.

Try piloting the changes on a single project – the results will be visible quickly and with rapid implementation, project teams can be up and running within a couple of weeks.

The big decision is not to run the pilot, as most organisations can manage to ring-fence a pilot project for its duration. The key moment comes once the pilot has proven to be successful. Given that it is not sustainable to run two different project management philosophies concurrently, once you have chosen to pilot, you should plan for its success, and for how this change will improve how you manage all future capex projects.

Pfeffer and Sutton show that you cannot rely on success alone to embed these practices into your organisation, as is demonstrated in the following examples

- Even before the concept of CCPM was made public in 1997, it was used in the UK construction sector. In the early 1990's Balfour Beatty established a Business Improvement Team, and between 1995 and 1997 they used CCPM on four trial projects. It was successful on all of them, and one project that used CCPM from start to finish delivered a project that was expected to take two years in under a year, maintaining the contractor's margin, despite absorbing a number of scope changes. Balfour Beatty's Business Improvement Team presented the results externally, claiming CCPM to be a great success. However, once the successful team was disbanded, the members went back to working for project directors who preferred to manage projects in their own way. They were under no pressure to change, so they didn't. As long as they produced expected (if not optimum) results, the board was happy. Since the share price was rising steadily in the early 2000's, there was no 'burning bridge' and nothing changed.

- In the mining industry, one of the global majors claimed great success using CCPM on a new production facility. However, this project wasn't on the board's radar, and changes in senior managers meant that the knowledge was lost, and sadly the same organisation is now struggling to develop new facilities as quickly and cost-effectively as their competition.

- The same story can be told with Project Alliancing. Despite its undoubted success in the UK's oil, gas, and chemical industries in the 1990's, Project Alliancing did not become the embedded norm. Although it has been used by a few individuals throughout their careers in the years since, by the time the oil price crashed in late 2014, the Oil and Gas industry globally had forgotten the lessons learned in the UK some 20 years earlier. A whole generation of Senior Executives had no awareness that their predecessors had developed, and achieved great results with, Project Alliancing.

- Even in Australia, after the successful application of alliances on many projects, there were powerful stakeholders who lobbied to return to more traditional forms of selection and contracting. The need to "blame", and the belief that fixed-price bids are accurate predictors of outturn costs, are very much alive and well.

We highlight these points, not to put people off, but to encourage you to take change seriously. The key point is that good ideas don't always become established just because they are good. It takes deliberate effort.

This Guide is just one step on the way. Our goal was to highlight why change is necessary, and to show a way in which you can achieve significant and sustainable improvement in the performance of your projects.

If you want to implement Breakthrough PM, we would suggest you address at least the following in your change programme.

- Start with a pilot to give you and your organisation confidence in the method. Not too easy a project, nor one with your A-team on it.

- Use internal specialist staff (or get external support) to support the early pilots, and ensure you correctly implement both CCPM and the Project Alliance.

- Isolate the pilot from business-as-usual, for example by not requiring them to produce your standard management reports.

- Get the board on board – ideally before you start. Give them awareness training so they can support the pilot. Report progress. Give the pilot visibility.

- Once you have enough evidence from the pilot, plan rollout as a major internal change project.

- Make this change holistic. In order to have a bottom-line impact, you will need to involve all functional areas of the business – from HR to sales, from marketing to legal, from training to QA.

  When your business adopts Breakthrough PM significant extra capacity is usually revealed, since you should be delivering the same projects with less resource. What will you do with this? Do you need additional sales to exploit it? Can you find other work for people to do? You need to make sure staff do not fear being made redundant.

Don't forget the team

Breakthrough PM relies on the project team working as a collaborative team. CCPM and Project Alliancing will remove the constraints to cross-company collaboration, and provide the team with a planning and execution method that both embeds and exploits collaboration.

However, the team will still require active leadership to ensure that the right behaviours and culture are embedded across the project team. And as well as leadership, the project team members all need the desire and capacity to follow, and to be true team members, as opposed to supplier there following orders.[51] This characteristic should be included in the criteria used to select alliance partners.

Leading and working in the project team has not been a major focus of this book, but that doesn't mean that this is not important.

Projects do not have long to develop the right culture and environment, and establishing the right culture needs to involve all the members of the project team, irrespective of how they are contracted. It is not something only for the members of the Project Alliance. Structured team-building events are usually much more productive with specialist facilitation, though most projects can manage team-socialising without much help. These are important, and

---

[51] There are several books on the topic of "followership" – see for example "*The Courageous Follower: Standing Up To and For Our Leaders*" by Ira Chaleff.

should be built into the project execution strategy and budget.

Implementation Examples

In the final section below, we describe how Breakthrough PM could be implemented in three different environments; the client, main contractors, and specialist sub-contractors.

The Investor/Client

For the client, the main issue is the lack of suppliers and contractors who are delivering projects in the way we outline. It is not as straightforward as simply adding CCPM and Project Alliancing into your tender documents.

In order to exploit Breakthrough PM and realise the significant increase in ROI that is possible, clients will need to change how they manage projects, and how they procure the services of project specialists. They also need to change how contractors manage their own business.

If the client is willing to be 'hands-on', they can employ project team specialists directly on a reimbursable basis, paying by the day, and use CCPM to plan and manage execution. This is the main route capex project clients have used when implementing CCPM in the past.

Breakthrough PM gives you another option. You can encourage contractors and suppliers to enter into a Project Alliance to deliver your project, and require them to use CCPM in the process.

This requires a slightly different approach to selecting your project team; it will require pre-selection communication with potential suppliers and for contractors to engage with your ideas. You will need to use methods of supplier development and reverse marketing[52] as much as supplier selection and appraisal.

Project Alliancing will probably be more familiar than CCPM to many contractors, and they will usually be quite willing to work under a Project Alliance. CCPM may need a little persuasion. Neither ought to be a stumbling block, but each will need some thought and preparation to ensure you give your project the best chance of success. This begins with choosing the keenest and most competent supply partners.

If you are selecting a single main contractor, rather than putting together a multi-party alliance, then it is important that the main contractor agrees to implement Breakthrough PM with their supply chain. It only works if waste is removed from the project, and since most main contractors subcontract 70-80% of the work, their major sub-contractors will also need to be integrated.

Typically, a pre-meeting with interested bidders, explaining the ideas of a Project Alliance and CCPM will be sufficient for them to be very interested when they receive your tender package. Of course this tender, and

---

[52] Reverse Marketing is about proactively persuading suppliers to offer goods/services that they do not currently provide. See *Reverse Marketing: The New Buyer-supplier Relationship*, by Leenders & Blenhorn, Macmillan USA, 1987.

the selection process, will have to be based on the ideas in this book.

During execution you will also need to ensure contractors are implementing Breakthrough PM correctly, and we recommend that you make training, coaching, and facilitation available to the project team. Any additional cost in establishing Breakthrough PM in your project will be more than offset by the reduced cost of project control (especially cost control) compared to a traditional project.

If you are a private sector organisation, we suggest that you begin with a discussion with your existing suppliers. You only need one who is willing to 'give it a go' and you are underway. You do not need to promise any long-term commitment since a Project Alliance delivers benefit to all parties on a single project, and your chosen contractors will be learning valuable lessons by working with you in this way. They will develop expertise and knowhow that could set their business apart from their competition just as much as it sets your business apart from yours.

Numerical Example

We simulated a £100 million speculative construction project, funded by interest-bearing debt, with an expected profit over a 13-year period of £50M (3-year construction, 10 years of sales and rental income). According to the plan, capital and interest was repaid early in year 10, more than 6 years after completion.

If this simulated project, like most projects, is late and over budget, the profit falls to zero. Capital and interest is eventually repaid in year 14.

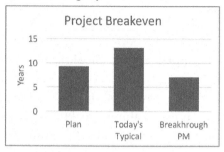

However, if the project follows the recommendations in this book:

- The project makes $89M profit over the 13 years – almost double the plan. This profit, if left in the business would pay for an additional project without the need for external funding. This is like going to the supermarket – ***buy one building, get another free!***[53]

- The debt and interest is paid off early in year 7. Assuming the developer has limited credit available, this equates to the capacity to increase the development portfolio by 25-35%, with the same lending and the same resources

- And these figures do not even take into account of the value of coming to market ahead of your competition[54], or selling your product at a premium price due to earlier completion, which may be of significant value to your clients

---

[53] In supermarkets the 'buy one, get on free' offer is known by the acronym BOGOF. We have invented the BOBGAF!
[54] One of the main fields that has used CCPM since its introduction in the 1990's has been new product development, where market opportunities for new technologies change very rapidly.

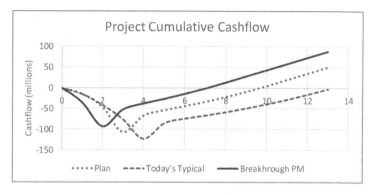

Not all clients invest in order to create future income. Some *have to* invest, for example to replace aging assets, or because of regulatory changes. Because Breakthrough PM delivers projects in a shorter duration, this can allow these type of projects to be started later, releasing resources and cash to do value-adding work in the meantime.

The Main Contractor

Even if the client does not require CCPM, or a Project Alliance, Breakthrough PM can be used by a main contractor (MC)[55] to put together their own project team, and to significantly improve the profitability of traditionally-bid contracts.

The MC could <u>pilot the idea on an existing contract</u>, at relatively low risk. The ideal project would be one where the client is relatively hands-off, and leaves the MC to manage the project and select most suppliers and sub-

---

[55] We are not going to worry about subtle differences between 'prime contractor', 'main contractor', 'general contractor', etc. For our purposes they are all the same – the client contracts with them to do the project (or most of it).

contractors themselves. It will probably be a lump-sum contract, because often with other forms, clients want visibility of, and a say in, supplier selection. The MC should have already identified which of their specialist sub-contractors would be most willing to try CCPM and capable of entering into a contract using a Project Alliance. The team would be selected early, and would sign up to the cost and time targets, based on the already signed main contract, and develop the CCPM schedule from there. Ideally the provisional Project Alliance team had already supported the MC's bid, and so will already have good understanding of the project and its objectives. It would then be over to the team to put it into action and to further exploit the collaborative project team.

The <u>strategic benefits however, only come when the method is rolled out across the wider business</u>.

Having completed a successful project ahead of the due-date and with more profit than with a traditional approach, the MC will now have the confidence in the method, and in particular that they can deliver contracts faster and for lower cost than they did before.

The MC is now able to:

- Sign-up to much higher liquidated damages or lateness penalties than their competitors, thus reducing competition on bids where time is important to the client. They could even encourage clients to include higher-than-normal damages/penalties in the tender, to scare off the 'suicide bidders',[56] thereby securing more work at normal, or higher, prices

- Increase their volume of business, without taking on more staff or overheads (remember that delivering projects in 75% of the traditional time means that the same resources can do 33% more work, without being overloaded). Many construction companies worry about growing too fast because this means new project teams, and a higher risk because you do not know how good the new people are. Breakthrough PM gives a way to grow without taking this risk.

- Reduce their bidding prices, confident that your time-related delivery costs will be lower. This means that the lower bid costs still make more profit for the MC and their supply chain than under the traditional regime.

Numerical Example

If we take a contractor, with 20% gross margin, overheads at 16%, and making a net profit of 4%, then:

- Increasing their turnover by 33%, without increasing their overhead and operating cost, would increase net profit by 2.7 times (to 11% return on sales);

- This assumes that they can win work at the same pricing as today. Even if they had to cut prices on the new business by 10% in order to secure this increase in volume, the net profit would still be 83% higher (at

---

[56] Suicide bidding is bidding a price so low that gross margin is very low, or even negative. The bidder just wants the turnover to try and keep their company together, and only after they have won do they worry about how to deliver the project, and how to make a profit.

8% ROS).

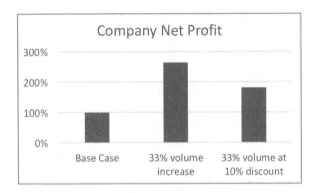

With proven references to call on to endorse the virtues of them using Breakthrough PM, the MC is also better suited to win more business, even where the client takes a more active role.

The Specialist Sub-contractor

Specialists who subcontract a major portion of their work (for example building services, civil works, and MEP (mechanical-electrical-plumbing), can benefit in the same way as the main contractors above; they can bid with a traditional fixed price, confident in the knowledge that their supply chain can deliver faster and at lower cost, whilst still making higher profits.

The benefits from faster delivery are more limited for the specialist sub-contractor. If the main contractor is not using CCPM or Breakthrough PM, it is unlikely that they (or their client) will make use of the earlier completion by the specialist. Even worse, the main contractor may deliberately slow the progress of the site team, and require the sub-contractor to be on-site longer than necessary.

Another significant problem is the main contractor and other sub-contractors may create pressures to multi-task and start tasks too early, just to look busy, and in the mistaken belief that 'the earlier you start the sooner you finish'. Interference from other contractors working on the project is another issue, particularly when each has an independent contract.

The specialist may be able to reduce project overhead (prelim)[57] costs, to deliver more work with the same resources, particularly if a significant portion of their work is remote from the main project team, or at least is relatively independent. Most of this benefit though will come from implementing CCPM, rather than through Project Alliancing.

The main opportunity for many specialist sub-contractors is to expand their scope of work into the overall project management and develop into a main contractor role. Many main contractors had their roots in a specific trade. Many have a civil/structural background, and stepped into a main contractor role simply because they were first on site, and therefore the first to be selected by the client.

Developing experience and a capability in the methods of Breakthrough PM may open new doors and opportunities; where – at least in the short term – there is much less competition. Which brings us to…

---

[57] Prelims, or preliminaries, are a term used in the construction sector to refer to project overhead costs such as the site establishment and the project management team.

The Specialist Sub-contractor Alliance

An interesting avenue that specialist sub-contractors might also consider is working together with other non-competing specialists, to form a 'virtual main contractor'.

This group creates a pre-formed Project Alliance, and bids for project contracts in competition with main contractors. This could be formalised as a joint-venture (JV) or limited company with each specialist being a shareholder, and a JV structure based on the principles of the Project Alliance. Alternatively, the organisation can be less formal, with one member can act as the main contracting entity on behalf of the group, and they immediately sub-contract the whole project to the partners under a Project Alliance.

Whichever way it is set up, the most important factor is how the team operates, rather than the legal structure. It will work just like a Project Alliance but without one single 'client'.

The ability to self-coordinate, and establish trust amongst the team, allows the group to deliver projects without the additional cost of an extra managerial layer between the specialist and the end client. And of course they can use CCPM to deliver the project (and more!) reliably.

One example of this idea is a US organisation called Integrated Project Delivery (http://ipdfl.net).

Integrated Project Delivery was formed by five partners in the construction field; an architect, a general contractor, a mechanical specialist, an electrical specialist, and a heating and ventilation specialist. They formed a legal entity to bid for whole projects on a design and build basis, working together using the principles of Project Alliancing and collaborative contracting.

So if you're a sub-contractor, frustrated being treated like a commodity by main contractors, who you don't believe will even consider using an innovation like Breakthrough PM, why not get together with like-minded other specialists and consider 'cutting out the middleman'?

End Note

We started this book, with a quotation attributed to Einstein…

> *"We cannot solve our problems with the same thinking we used when we created them."*

Over the previous pages we have outlined why we feel that capex projects have problems, and have shared how we believe these problems can be overcome, with different thinking, and different methods.

We hope the knowledge shared in this book will help you achieve significant improvements, and look forward to hearing about your experiences.

If you want to find out more about Breakthrough Project Management, access a range of additional support services and information, and to join our community of change agents, visit our website:

www.breakthroughprojectmanagement.com

We look forward to seeing you there.

Bibliography & References

There is a wealth of resources available that support the ideas and methodologies in this book. Here we list a brief selection to help interested readers begin to research the topics in more detail.

These books, journals, websites and articles are a great place to find out more about CCPM, Collaborative Contracting, Project Alliances, and teamwork and we hope you find exploring them interesting and informative.

Critical Chain Project Management

Author	Year	Title & Details
Goldratt, EM	1997	*Critical Chain* North River Press
Kendal, G & Austin, K	2012	*Advanced Multi-Project Management* J Ross Publishing
Kishira, Y	2009	*Wa: Transformation Management by Harmony.* North River Press
Leach, L	2014	*Critical Chain Project Management*, Third Edition Artech House
Newbold, R	2011	*Billion Dollar Solution:* Secrets of ProChain Project Management. ProChain Solutions Inc

Project Alliancing & Commercial Partnering

Author	Year	Title & Details
American Institute of Architects	2007	*Integrated Project Delivery: A Guide* http://www.aia.org/contractdocs/aias077630
CII	1996	*Model for Partnering Excellence*, Research Summary 102-1. The Construction Industry Institute, University of Texas at Austin

CRINE	1994	*The CRINE Report:* Cost Reduction Initiative for the New Era. LOGIC, Oil & Gas UK
Government of Australia	2011	*National Alliance Contracting Guidelines*[58] https://infrastructure.gov.au/infrastructure/ngpd/files/National_Guide_to_Alliance_Contracting.pdf
Jones, D	2012	*Relationship Contracting* Chapter 3 in *The Projects and Construction Review*, 2nd Edition, Editor Júlio César Bueno, Law Business Research Ltd, London.
Ross, J	2003	*Introduction to Project Alliancing (on engineering & construction projects).* April 2003 update. PCI Group, Australia. http://www.pcigroup.com.au/publications_pci/
State of Victoria	2009	*In Pursuit of Additional Value.* A benchmarking study into alliancing in the Australian Public Sector. http://www.dtf.vic.gov.au/Publications/Infrastructure-Delivery-publications/In-pursuit-of-additional-value
Vitasek, K, & Manrodt, K	2012	*The Vested Way* Palgrave Macmillan
Yeung, Chan & Chan	2012	*Defining relational contracting from the Wittgenstein family-resemblance philosophy* International Journal of Project

[58] There are some aspects of the implementation of project alliancing in Australia that we do not fully agree with, and it is structured to the requirements of the public sector.

Management, February 2012

CapEx Project Management

Author	Year	Title & Details
AT Kearney	2012	*ExCap II: Top Level Thinking on Capital Projects* www.atkearney.com
CRINE	1994	*The CRINE Report:* Cost Reduction Initiative for the New Era. LOGIC, Oil & Gas UK
Egan J. et al	1998	*Rethinking Construction*: Report of the Construction Task Force ("The Egan Report") HMSO, London
EY (Ernst & Young)	2014	*Spotlight on oil and gas megaprojects* www.ey.com
KPMG	2013	*Avoiding Major Project Failure – Turning Black Swans White* www.kpmg.com
Latham M. et al	1994	*Constructing the Team* ("The Latham Report") HMSO, London
Lean Construction Institute	2014	*Construction Productivity in Decline* http://www.leanconstruction.org/media/docs/PEJune14_Construction.pdf
McKinsey	2013	*Infrastructure productivity: how to save $1 trillion a year* www.mckinsey.com/mgi

| McKinsey | 2015 | *The construction productivity imperative.* http://www.mckinsey.com/industries/infrastructure/our-insights/the-construction-productivity-imperative |

Collaboration & Teamwork

Author	Year	Title & Details
Braton, W & Tumin, Z	2012	*Collaborate or Perish* Crown Business, a division of Random House.
Collins, J	2001	*Good to Great* Random House Business
Hefferman M	2014	*A Bigger Prize:* Why no one wins unless everybody wins. Simon & Schuster UK.
Lencioni, P	2002	*The Five Dysfunctions of a Team* John Wiley & Sons
Pfeffer, J & Sutton, R	1999	*The Knowing-Doing Gap: How Smart Companies Turn Knowledge into Action* Harvard Business Review Press
Sawyer, K	2007	*Group Genius:* The Creative Power of Collaboration. Basic Books, New York

The Authors

Ian and Robert have over 50 years of experience on capex projects, working for consultants, contractors and clients.

As engineering graduates, they both started their careers working on projects; Robert with one of Australian's largest construction contractors, and Ian on the client-side with one of the world's largest chemical companies.

Their consulting experience includes working for major players such as PWC, along with niche specialists including Proudfoot/Crosby, REL/Hackett, PMMS/ArcBlue, Newport, and Pinnacle Strategies.

In the 1990's they both came across CCPM, with Robert working directly with Dr Eli Goldratt's team during its development. It was at this time that they both saw its potential to improve capex project performance. Their careers then took them outside the industry, and they were intrigued as to why CCPM did not gain the prominence it deserved.

The initial idea of writing this book came in the summer of 2014, when they were discussing the low levels of CCPM use on capex projects - at 30,000 feet, on a flight from Washington to Dubai.

Ian Heptinstall

ian@BreakthroughProjectManagement.com

In his early career, Ian managed projects in the process industries in the UK, France and Belgium, followed by time as a project management coach and advisor.

In the late 1990's he held a lead role in an award-winning project that was one of the first to apply the Project Alliance principles developed in the Oil and Gas industry's CRINE initiative to smaller projects outside of the Oil and Gas sector.

Around 2000, he moved into a global procurement role in the pharmaceutical industry, and later that decade was Chief Procurement Officer for a UK construction company before moving into full-time consultancy in 2011.

As a consultant Ian travels frequently to work with clients across the globe, from Japan to the US, and extensively in the Middle East, Africa and Europe.

Ian is a qualified mechanical engineer, and Fellow of the Chartered Institute of Procurement and Supply. He lives in the UK and Switzerland.

Robert Bolton

robert@BreakthroughProjectManagement.com

After graduating in civil engineering, Robert commenced his career in the construction industry. He has managed large capex projects in the civil, building and mining sectors in Australia.

His consultancy endeavours continued his work in the complex projects field, with extensive experience in the mining and capital markets sectors, in addition to more general business and manufacturing improvement.

He is a pioneer in Critical Chain Project Management (CCPM), having been involved in its conception and early development in the 1990's.

More recently he was involved in a major Oil & Gas project rescue, working at client locations in Singapore, Malaysia and China.

Like Ian, as a consultant he works with clients across the globe to help speed up their projects and business processes.

Robert has an MBA from Ashridge, and lives in Australia.

CPSIA information can be obtained
at www.ICGtesting.com
Printed in the USA
BVOW11s1128050617
5632BVAU00001B/1/P